APPROXIMATING COUNTABLE MARKOV CHAINS

HOLDEN-DAY SERIES IN PROBABILITY AND STATISTICS

E. L. Lehmann, Editor

Approximating Countable Markov Chains
David Freedman

Brownian Motion and Diffusion
David Freedman

Markov Chains
David Freedman

Nonparametric Statistics
Jaroslav Hajek

Basic Concepts of Probability and Statistics
J. L. Hodges, Jr., and E. L. Lehmann

Elements of Finite Probability
J. L. Hodges, Jr., and E. L. Lehmann

Mathematical Foundations of the Calculus of Probability
Jacques Neveu

Stochastic Processes
Emanuel Parzen

Foundations of Probability
Alfred Renyi

APPROXIMATING COUNTABLE MARKOV CHAINS

DAVID FREEDMAN
University of California, Berkeley

HOLDEN-DAY

San Francisco

TO WILLIAM FELLER

QA
274.7
.F73

PREFACE

A long time ago I started writing a book about Markov chains, Brownian motion, and diffusion. I soon had two hundred pages of manuscript and my publisher was enthusiastic. Some years and several drafts later, I had a thousand pages of manuscript, and my publisher was less enthusiastic. So we made it a trilogy:

Markov Chains
Brownian Motion and Diffusion
Approximating Countable Markov Chains

familiarly — *MC, B & D,* and *ACM.*

I wrote the first two books for beginning graduate students with some knowledge of probability; if you can follow Sections 10.4 to 10.9 of *Markov Chains,* you're in. The first two books are quite independent of one another, and completely independent of this one, which is a monograph explaining one way to think about chains with instantaneous states. The results here are supposed to be new, except when there are specific disclaimers. It's written in the framework of Markov chains; we wanted to reprint in this volume the *MC* chapters needed for reference, but this proved impossible.

Most of the proofs in the trilogy are new, and I tried hard to make them explicit. The old ones were often elegant, but I seldom saw what made them go. With my own, I can sometimes show you why things work. And, as I will argue in a minute, my demonstrations are easier technically. If I wrote them down well enough, you may come to agree.

The approach in all three books is constructive: I did not use the notion of separability for stochastic processes and in general avoided the uncountable axiom of choice. Separability is a great idea for dealing with any really large class of processes. For Markov chains I find it less satisfactory. To begin with, a theorem on Markov chains typically amounts to a statement about a probability on a Borel σ-field. It's a shame to have the proof depend on the existence of an unnamable set. Also, separability proofs usually have two parts. There is an abstract part which establishes the existence of a separable version. And there is a combinatorial argument, which establishes some property of the separable version by looking at the behavior of the process on a countable set of times. If you take the constructive approach, the combinatorial argument alone is enough proof.

When I started writing, I believed in regular conditional distributions. To me they're natural and intuitive objects, and the first draft was full of them. I told it like it was, and if the details were a little hard to supply, that was the reader's problem. Eventually I got tired of writing a book intelligible only to me. And I came to believe that in most proofs, the main point is estimating a probability number: the fewer complicated intermediaries, the better. So I switched to computing integrals by Fubini. This is a more powerful technique than you might think and it makes for proofs that can be checked. Virtually all the conditional distributions were banished to the Appendix. The major exception is Chapter 4 of *Markov Chains,* where the vividness of the conditional distribution language compensates for its technical difficulty.

In *Markov Chains,* Chapters 3 to 6 and 8 cover material not usually available in textbooks — for instance: invariance principles for functionals of a Markov chain; Kolmogorov's inequality on the concentration function; the boundary, with examples; and the construction of a variety of continuous-time chains from their jump processes and holding times. Some of these constructions are part of the folklore, but I think this is the first careful public treatment.

Brownian Motion and Diffusion dispenses with most of the customary transform apparatus, again for the sake of computing probability numbers more directly. The chapter on Brownian motion emphasizes topics which haven't had much textbook coverage, like square variation, the reflection principle, and the invariance principle. The chapter on diffusion shows how to obtain the process from Brownian motion by changing time.

I studied with the great men for a time, and saw what they did. The trilogy is what I learned. All I can add is my recommendation that you buy at least one copy of each book.

User's guide to *Approximating Countable Markov Chains*

I've covered Chapters 1 and 3 in a seminar with advanced graduate students; it took about fifty sessions. I've never exposed Chapter 2 to scrutiny, but I think you could do it in about twenty sessions.

Acknowledgments

Much of the trilogy is an exposition of the work of other mathematicians, who sometimes get explicit credit for their ideas. Writing *Markov Chains* would have been impossible without constant reference to Chung (1960). Doob (1953) and Feller (1968) were also heavy involuntary contributors. The diffusion part of *Brownian Motion and Diffusion* is a peasant's version of Itô and McKean (1965).

The influence of David Blackwell, Lester Dubins and Roger Purves will be found on many pages, as will that of my honored teacher, William Feller. Ronald Pyke and Harry Reuter read large parts of the manuscript and made an uncomfortably large number of excellent suggestions, many of which I was forced to accept. I also tested drafts on several generations of graduate students, who were patient, encouraging and helpful. These drafts were faithfully typed from the cuneiform by Gail Salo.

The Sloan Foundation and the US Air Force Office of Scientific Research supported me for various periods, always generously, while I did the writing. I finished two drafts while visiting the Hebrew University in Jerusalem, Imperial College in London, and the University of Tel Aviv. 1 am grateful to the firm of Cohen, Leithman, Kaufman, Yarosky and Fish, criminal lawyers and xerographers in Montreal. And I am still nostalgic for Cohen's Bar in Jerusalem, the caravansary where I wrote the first final draft of *Approximating Countable Markov Chains*.

David Freedman

Berkeley, California
May, 1972

TABLE OF CONTENTS

1. RESTRICTING THE RANGE

1. Summary of chapters 1 and 2	1
2. Inequalities	4
3. Standard transitions	8
4. Recurrence	10
5. Restricting the range	12
6. The Markov property	18
7. The convergence of X_J to X	20
8. The distribution of γ_J given X_J	25
9. The joint distribution of $\{X_J\}$	30
10. The convergence of Q_J to Q	43
11. The distribution of X given X_J	51

2. RESTRICTING THE RANGE; APPLICATIONS

1. Foreword	64
2. The generator	64
3. A theorem of Lévy	68
4. Determining the time scale	71
5. A theorem of Williams	77
6. Transformation of time	78
7. The transient case	89

3. CONSTRUCTING THE GENERAL MARKOV CHAIN

1. Introduction	95
2. The construction	96
3. A process with all states instantaneous and no pseudo-jumps	111
4. An example of Kolmogorov	114
5. Slow convergence	116
6. Smith's phenomenon	119

4. APPENDIX

   1. Notation   128

   2. Numbering   129

   3. Bibliography   129

BIBLIOGRAPHY   131

INDEX   137

SYMBOL FINDER   139

APPROXIMATING
COUNTABLE
MARKOV CHAINS

1

RESTRICTING THE RANGE

1. SUMMARY OF CHAPTERS 1 AND 2

Let X be a Markov chain with state space I, stationary standard transitions P, and smooth sample functions. For simplicity, suppose I forms one recurrent class of states relative to P. Let J be a finite subset of I. Let X_J be X watched only when in J; namely, X_J is obtained from X by ignoring the times t with $X(t) \notin J$. Call X_J the *restriction* of X to J. This operation has been considered by Lévy (1951, 1952, 1958) and Williams (1966). The process X_J is defined formally in Section 5. The idea is to introduce $\gamma_J(t)$, the rightmost time on the X-scale to the left of which X spends time t in J. Then $X_J(t) = X[\gamma_J(t)]$. Suppose J and K are finite subsets of I, with $J \subset K$. Then $X_J = (X_K)_J$, as shown in Section 5. The process X_J is Markov with stationary standard transitions, say P_J, on J. This is proved in Section 6. The semigroups $\{P_J : J \subset I\}$ are equicontinuous; consequently, X_J converges to X in probability and in q-lim† with probability 1 as J increases to I; in particular, P_J converges to P as J increases to I. These results are proved in Section 7.

The distribution of the process γ_J is studied in Section 8. To state the main result, let τ_J be the first holding time in X_J. Let α_J be the time X spends interior to the first interval of constancy for X_J: so $\alpha_J = \gamma_J(\tau_J -)$. Suppose $X(0)$ is fixed at $i \in J$. Then α_J is independent of $X(\alpha_J + \cdot)$ and is distributed like $G(\tau_J)$, where:

(a) the process G is independent of X;

(b) G has stationary, independent increments;

(c) $G(0) = 0$;

(d) $\lim_{t \downarrow 0} G(t)/t = 1$;

(e) $t \to G(t) - t$ is right-continuous and nondecreasing.

Of course, G too depends on J.

†For the definition, see page 302 of *MC*.

Suppose K consists of J and one additional element k. The conditional distribution of X_K given X_J is obtained in Section 9. To describe it, let $Q_J = P'_J(0)$ and $q_J(i) = -Q_J(i, i)$. Then X_K is obtained by cutting X_J and inserting k-intervals. Given X_J and the locations of the cuts, the lengths of the inserted intervals are independent and exponential, with common parameter $q_K(k)$. There are two kinds of cuts: the first kind appears at a jump of X_J, and the second kind appears interior to an interval of constancy for X_J. Cuts of the first kind appear independently from jump to jump, and independently of cuts of the second kind. Locations of cuts of the second kind are independent from interval to interval. At a jump from i to j, the probability of a cut not appearing is $Q_K(i, j)/Q_J(i, j)$. Within a particular j-interval, the location of cuts has a Poisson distribution, with parameter $q_K(j) - q_J(j)$.

One of the main results of this chapter is proved in Section 10, namely: $P'_J(0)$ is nondecreasing and tends to $P'(0)$ as J increases to I. In fact, for each $i \neq j$ in I and positive ε, there is a positive δ such that for all $t \leq \delta$ and finite $J \supset \{i, j\}$,

(a) $$P_J(t, i, j) \geq (1 - \varepsilon)P(t, i, j).$$

Moreover, there is a finite J with $\{i, j\} \subset J$ and a positive δ such that for all $t \leq \delta$ and finite $K \supset J$,

(b) $$P_K(t, i, j) \leq (1 + \varepsilon)P(t, i, j) + \varepsilon t.$$

If $Q(i, j) > 0$, the second term on the right in (b) is plainly redundant. If $Q(i, j) = 0$, the first term on the right in (b) is redundant, but the second term is vital: for $Q_K(i, j)$ is likely to be positive. The analog for $i = j$ fails:

$$(1 - \varepsilon)[1 - P(t, i, j)] \leq 1 - P_J(t, i, i) \quad \text{for } t \leq 1/17$$

is impossible when i is instantaneous.

The conditional distribution of X given X_J is studied in Section 11. Remember that τ_J is the first holding time in X_J; while α_J is the time X spends interior to the first interval of constancy for X_J. Let

$$Y(t) = X(t) \quad \text{for } 0 \leq t < \alpha_J.$$

Fix $X(0)$ at i in J. Then Y is independent of $X(\alpha_J + \cdot)$. And Y is distributed like

$$\{Y^*(t) : 0 \leq t < \lambda(\tau_J)\},$$

where the process Y^* is independent of X, and Markov with starting state i and stationary transitions; while $\lambda(t)$ is the least time to the left of which Y^* has spent time t in state i. Of course, Y^* too depends on J.

Sections 2, 3, and 4 contain some lemmas with wider application. The results of Section 2 are taken from (Blackwell and Freedman, 1968). To state the main ones, let P be a standard stochastic semigroup on I; let $i \in I$; let

$$f(t) = P(t, i, i);$$

and let

$$m = \min \{f(t) : 0 \leq t \leq 1\}.$$

Then

$$\frac{1 + m}{2} > \int_0^1 f(t) \, dt,$$

provided the integral exceeds $\frac{3}{4}$. Furthermore,

$$m \geq \frac{1 + \sqrt{4f(1) - 3}}{2},$$

provided $f(1) > \frac{3}{4}$.

The theorem in Section 3 appears in (Chung, 1960, p. 141), and in a more general context in (Kingman, 1968). It says that the transitions of a measurable Markov chain which visits each state with positive mean time are necessarily standard. Section 4 is routine; for another treatment, see (Chung, 1960, II.10). It concerns the partition of the states into recurrent classes.

The results of Chapter 1 are used in Chapter 2. One of the main results in that chapter is (2.2), which provides an interpretation of $Q = P'(0)$ in terms of the sample function behavior of X. For simplicity, suppose I forms one recurrent class of states relative to P. Fix $i \in I$ and suppose X starts from i. For $j \neq i$ in I, let $\theta(j)$ be the time X spends in i until the first pseudo-jump from i to j: where X pseudo-jumps from i to j at time σ iff $X(r) = j$ for binary rational r arbitrarily close to σ on the right, while $X(s) = i$ for binary rational s arbitrarily close to σ on the left. The $\theta(j)$ are independent as j varies over $I \setminus \{i\}$, and $\theta(j)$ is exponentially distributed with parameter $Q(i, j)$. The proof of (2.2) uses Section 1.10. Lévy's theorem, that $P(t, i, j)$ is positive everywhere or nowhere on $(0, \infty)$, is proved in Section 2.3. This proof uses Section 1.8, but not the later sections of Chapter 1. Suppose X is subjected to a homogeneous change of time scale $t \rightarrow t/c$. Then c can be computed with probability 1 from a piece of sample function iff the piece is not a step function. This result is proved in Section 2.4, using only Section 6 from Chapter 1. A theorem of Williams is proved in Section 2.5. Fix $i \in I$. If $P'(0, i, j) \geq \varepsilon > 0$ for all $j \neq i$, then i is instantaneous and all other states are stable. The proof uses Section 1.9 and (1.88).

Section 2.6 characterizes all the processes which can be obtained from X through certain inhomogeneous changes of time; this uses Section 1.7, but no later sections of Chapter 1. To state the result, suppose I forms one recurrent class of states. For finite $J \subset I$, let λ_J be the least t with $X(t) \in J$; and let $P(i, J)$ be the P_i-distribution of $X(\lambda_J)$, a probability on J. Suppose P and R are standard stochastic semigroups on I, with $P(i, J) = R(i, J)$ for all i and J. Then there is a unique positive function f on I, such that $M_f(t)$ is finite for all t and X_f is Markov with stationary transitions R relative to P_i, where:

$$M_f(t) = \int_0^t f[X(s)] \, ds$$

$$T_f = M_f^{-1}$$

$$X_f = X \circ M_f.$$

All the preceding theory, in Chapters 1 and 2, is developed only for recurrent chains. Section 2.7 indicates the modifications needed to handle the general case.

2. INEQUALITIES

Let P be a standard stochastic semigroup on the countable set I. Fix one state $a \in I$ and abbreviate

$$f(t) = P(t, a, a).$$

(1) Theorem.   *Suppose $0 < \varepsilon < 1$ and $f(1) \leq 1 - \varepsilon$. Then*

$$\int_0^1 f(t) \, dt < 1 - \tfrac{1}{2}\varepsilon.$$

(2) Theorem.   *Suppose $0 < \varepsilon < \tfrac{1}{4}$ and $f(1) \geq 1 - \varepsilon$. Then for all t in $[0, 1]$,*

$$f(t) \geq \frac{1 + \sqrt{1 - 4\varepsilon}}{2} = 1 - \varepsilon - \varepsilon^2 - O(\varepsilon^3) \quad as \; \varepsilon \to 0.$$

(3) Corollary.   *If $0 < \delta < \tfrac{1}{2}$ and $f(1) \geq 1 - \delta + \delta^2$, then $f(t) \geq 1 - \delta$ for all t in $[0, 1]$.*

(4) Theorem.   *Suppose $0 < \varepsilon < \tfrac{1}{4}$ and $\int_0^1 f(t) \, dt \geq 1 - \varepsilon$. Then $f(t) > 1 - 2\varepsilon$ for all t in $[0, 1]$.*

It is worth noting explicitly that these bounds hold for all standard stochastic semigroups P and states a. They are not sharp, but cannot be improved much; compare examples $(MC, 8.35)$, $(MC, 8.36)$, and (16) below.

Theorem (1) asserts

$$\frac{1 + f(1)}{2} > \int_0^1 f(t)\, dt.$$

Theorem (4) asserts

$$\frac{1 + m}{2} > \int_0^1 f(t)\, dt,$$

where

$$m = \min\{f(t) : 0 \le t \le 1\};$$

provided the integral exceeds $\frac{3}{4}$. In this range, (4) is clearly sharper; I do not know whether (4) holds for all values of the integral. Theorem (2) asserts

$$m \ge \frac{1 + \sqrt{4f(1) - 3}}{2},$$

provided $f(1) > \frac{3}{4}$.

PROOF OF (1). Suppose a is not absorbing, so

(5) $0 < f(t) < 1$ for $0 < t \le 1$.

Suppose

(6) $f(1) = 1 - \delta.$

Plainly,

(7) $f(t) \cdot f(1 - t) \le f(1)$ for $0 \le t \le 1$.

A result slightly sharper than (1) will be proved:

(8) Proposition. *If f is measurable on $[0, 1]$ and satisfies (5), (6) and (7), then*

$$\int_0^1 f(t)\, dt < 1 - \tfrac{1}{2}\delta.$$

Indeed, if $0 < x < 1$ and $0 < y < 1$, then

$$0 < (1 - x)(1 - y) = 1 - x - y + xy$$

and $x + y < 1 + xy$. Consequently,

$$f(t) + f(1 - t) < 2 - \delta   \text{ for } 0 < t < 1.$$

Integrate from 0 to 1 and divide by 2.   ★

Harry Reuter eliminated the unnecessary part of an earlier proof of (8).

PROOF OF (2).   For $0 < t < 1$,

$$f(1) = f(t)f(1 - t) + \Sigma_{j \neq a} P(t, a, j)P(1 - t, j, a)$$
$$\leq f(t)f(1 - t) + \Sigma_{j \neq a} P(t, a, j).$$

Thus

(9) $f(1) \leq f(t)f(1 - t) + 1 - f(t)$ *for* $0 \leq t \leq 1$.

A result slightly sharper than (2) will be proved:

(10) Proposition.   *Suppose* f *is continuous on* $[0, 1]$. *Suppose* $f(0) = 1$ *and* $0 \leq f \leq 1$ *and* $f(1) \geq 1 - \varepsilon$, *where* $0 < \varepsilon < \frac{1}{4}$. *Suppose* f *satisfies* (9). *Then*

$$f(t) \geq \frac{1 + \sqrt{1 - 4\varepsilon}}{2} \quad \text{for } 0 \leq t \leq 1.$$

Begin the proof of (10) by establishing

(11) $f(t) \geq 1 - \sqrt{\varepsilon}$ for $0 \leq t \leq 1$.

Use (9) and the condition $f(1) \geq 1 - \varepsilon$:

(12) $f(t)[1 - f(1 - t)] \leq \varepsilon$ for $0 \leq t \leq 1$.

If $x, y \geq 0$ and $xy \leq \varepsilon$, then $x \leq \sqrt{\varepsilon}$ or $y \leq \sqrt{\varepsilon}$. Thus,

(13)   for each t in $[0, 1]$, either $f(t) \leq \sqrt{\varepsilon}$ or $f(1 - t) \geq 1 - \sqrt{\varepsilon}$.

Put $1 - t$ for t in (13) to get

(14)   for each t in $[0, 1]$, either $f(t) \geq 1 - \sqrt{\varepsilon}$ or $f(1 - t) \leq \sqrt{\varepsilon}$.

Now suppose $f(t) > \sqrt{\varepsilon}$, for a specific t in $[0, 1]$. Relation (13) implies $f(1 - t) \geq 1 - \sqrt{\varepsilon} > \sqrt{\varepsilon}$, because $\varepsilon < \frac{1}{4}$. Then relation (14) implies $f(t) \geq 1 - \sqrt{\varepsilon}$. That is,

(15)   for each t in $[0, 1]$, either $f(t) \leq \sqrt{\varepsilon}$ or $f(1 - t) \geq 1 - \sqrt{\varepsilon}$.

An easy continuity argument completes the proof of (11).

Introduce $\theta(x) = 1 - (\varepsilon/x)$, as in Figure 1. Let $b_0 = 1 - \sqrt{\varepsilon}$ and $b_{n+1} = \theta(b_n)$. If $f(t) \geq b_n$ for all t in $[0, 1]$, use (12) with $t = 1 - s$ to check that $f(s) \geq b_{n+1}$ for all s in $[0, 1]$. By algebra, θ is convex and has fixed points

$$b_{\pm} = \frac{1 \pm \sqrt{1 - 4\varepsilon}}{2}.$$

Moreover, $b_- < b_0 < b_+$. Hence $b_n \uparrow b_+$.   ★

PROOF OF (3).   Put $\varepsilon = \delta - \delta^2$ in (2).   ★

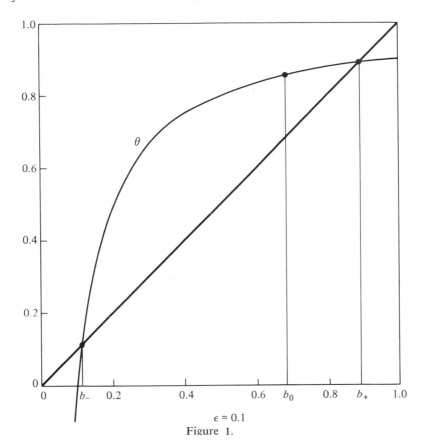

$\epsilon = 0.1$

Figure 1.

PROOF OF (4). By way of contradiction, suppose $0 < \varepsilon < \frac{1}{4}$ and $\int_0^1 f(s)\,ds \geq 1 - \varepsilon$ and $f(t) = 1 - 2\varepsilon$ for some t in $(0, 1]$. By (1),

$$\int_0^t f(s)\,ds < t(1 - \varepsilon).$$

Moreover, $2\varepsilon < \frac{1}{2}$, so (3) implies

$$f(s) \leq 1 - 2\varepsilon + 4\varepsilon^2 \quad \text{for } t \leq s \leq 1.$$

Consequently,

$$\int_t^1 f(s)\,ds \leq (1 - t)(1 - 2\varepsilon + 4\varepsilon^2)$$

$$= (1 - t)(1 - \varepsilon) - (1 - t)(\varepsilon - 4\varepsilon^2).$$

Add to get

$$\int_0^1 f(s)\, ds < (1 - \varepsilon) - (1 - t)(\varepsilon - 4\varepsilon^2).$$

But $\varepsilon - 4\varepsilon^2 > 0.$ ★

(16) Example. Let $q > 0$. Consider a Markov process with state space $\{0, 1\}$, such that the holding time parameter in 0 is q, and 1 is absorbing. Then

$$f(t) = P(t, 0, 0) = e^{-qt}.$$

As q tends to 0:

$$f(1) = e^{-q} = 1 - q + O(q^2);$$

$$\int_0^1 f(t)\, dt = (1 - e^{-q})/q = 1 - \tfrac{1}{2}q + O(q^2).$$ ★

QUESTION. Suppose $f(1)$ is known. Can you find sharp upper and lower bounds on $f(1/2)$? On $\int_0^1 f(t)\, dt$?

3. STANDARD TRANSITIONS

Let I be a finite or countably infinite set with the discrete topology. For finite I, let $\bar{I} = I$. For infinite I, let $\bar{I} = I \cup \{\varphi\}$ be the one-point compactification of I. Let $P(t, i, j)$ be a stochastic semigroup on I. Do not assume continuity. Suppose that for each $i \in I$, there is a probability triple (Ω_i, P_i), and a process X_i on (Ω_i, P_i) which is jointly measurable, $\bar{I}$-valued, and Markov with transitions P, starting from i. By Fubini, $P(t, i, j) = P_i\{X_i(t) = j\}$ is measurable in t. Suppose that for all i,

$$\int_0^\infty P(t, i, i)\, dt > 0.$$

NOTE. If X has quasiregular sample functions, then X is jointly measurable, as in $(MC, 9.15)$. *Quasiregular* is defined in $(MC, 9.7)$.

(17) Theorem. *P is standard.*

PROOF. Let $\Omega = \cup_i \Omega_i$. Define P_i on Ω by setting $P_i(\Omega_j) = 0$ for $j \neq i$. Define X on Ω by setting $X = X_i$ on Ω_i. Thus, with respect to P_i, the process X is Markov with transitions P starting from i, and X is jointly measurable $\bar{I}$-valued.

Fix $i \in I$. The problem is to show $\lim_{h \to 0} P(h, i, i) = 1$. Let

$$\mu(j) = \int_0^\infty P(t, i, j)\, dt,$$

and begin by assuming

(18) $\mu(j) < \infty$ for all $j \in I$.

Then verify

(19) $\Sigma_{j \in I}\, \mu(j)P(t, j, k) \leq \mu(k)$ for all $k \in I$.

Let $\mu = \Sigma_{j \in I}\, \mu(j)P_j$, a σ-finite measure on Ω which is not finite. Let t and s be non-negative, and let $j \in I$. I say

(20) $\mu\{X(t + s) = j\} \leq \mu\{X(t) = j\}$.

Indeed,

$$\mu\{X(t + s) = j\} = \Sigma_{a \in I}\, \mu(a)P(t + s, a, j)$$
$$= \Sigma_{a \in I}\, \Sigma_{b \in I}\, \mu(a)P(s, a, b)P(t, b, j)$$
$$= \Sigma_{b \in I}\, \Sigma_{a \in I}\, \mu(a)P(s, a, b)P(t, b, j)$$
$$\leq \Sigma_{b \in I}\, \mu(b)P(t, b, j) \qquad \text{by (19)}$$
$$= \mu\{X(t) = j\}.$$

Find $T > 0$ so $P(T, i, i) > 0$. Let

$$f(t, \omega) = 1 \quad \text{according as } X(t, \omega) = i$$
$$= 0 \qquad\qquad\qquad\qquad\quad \neq i.$$

Then f is jointly measurable. By $(MC, 10.58)$: as h tends to 0, so does

$$\int_0^T |f(t + h, \omega) - f(t, \omega)|\, dt.$$

For $h \leq 1$, this expression is at most twice

$$\int_0^{T+1} f(t, \omega)\, dt,$$

whose μ-integral is

$$\int_0^{T+1} \Sigma_{j \in I}\, \mu(j)P(t, j, i)\, dt \leq \int_0^{T+1} \mu(i)\, dt < \infty$$

by Fubini and (19). By Fubini and dominated convergence,

(21) $\int_0^T \int_\Omega |f(t + h, \omega) - f(t, \omega)|\, \mu(d\omega)\, dt \to 0.$

The inner integral in (21) is

$$\mu\{[X(t) = i \text{ and } X(t + h) \neq i] \cup [X(t) \neq i \text{ and } X(t + h) = i]\}$$
$$\geq \mu\{X(t) = i \text{ and } X(t + h) \neq i\}$$
$$= \mu\{X(t) = i\} \cdot [1 - P(h, i, i)]$$
$$\geq \mu\{X(T) = i\} \cdot [1 - P(h, i, i)] \qquad \text{by (20)}$$
$$\geq \mu(i)P(T, i, i)[1 - P(h, i, i)].$$

So (21) implies $P(h, i, i) \to 1$ as $h \to 0$.

To remove condition (18), apply the argument to the semigroup

$$t \to e^{-t} P(t).$$

That is, construct on Ω a random variable τ independent of X and exponential with parameter 1 for all P_i. If necessary, replace Ω by its Cartesian product with the Lebesgue unit interval. Let $\partial \notin \bar{I}$ and

$$Y(t) = X(t) \quad \text{for } t < \tau$$
$$= \partial \qquad \text{for } t \geq \tau.$$

Then Y has transitions $R(t) = e^{-t} P(t)$, extended to be absorbing at ∂, and Y is jointly measurable $\bar{I} \cup \{\partial\}$-valued. Let $v(j) = \int_0^\infty R(t, i, j) \, dt < \infty$ for $j \in I$, and let $v(\partial) = 0$. Verify that (19) holds for R and v, with $k \in I$; this relation is false for $k = \partial$. Let $\mathbf{v} = \Sigma \, v(i)P_i$. Verify (20) for $\mathbf{v}$ and Y, again with k in I, so not equal to ∂. The rest of the argument is the same, and proves

$$\lim_{h \to 0} e^{-h} P(h, i, i) = 1,$$

so

$$\lim_{h \to 0} P(h, i, i) = 1. \qquad \bigstar$$

NOTE. According to a theorem of Ornstein (Chung, 1960, p. 121), the function $P(\cdot, i, j)$ is identically 0 or strictly positive on $(0, \infty)$. So

$$\int_0^\infty P(t, i, i) \, dt > 0$$

is equivalent to

$$P(t, i, i) > 0 \quad \text{for some } t > 0.$$

4.   RECURRENCE

Let P be a standard stochastic semigroup on the finite or countably infinite set I. Fix i in I. By $(MC, 9.16)$, the Markov process X on the probability triple (Ω_q, P_i) has quasiregular† sample functions, starts from i, and has stationary transitions P. Let τ be the least $t \geq 1$ if any with $X(t) = i$, and $\tau = \infty$ if none. Let $f_i = P_i\{\tau < \infty\}$. Recall $S_i = \{t : X(t) = i\}$.

(22) Theorem. **(a)** *If $f_i < 1$, then S_i is bounded with P_i-probability 1 and $\int_0^\infty P(t, i, i)\, dt < \infty$.*

(b) *If $f_i = 1$, then Lebesgue $S_i = \infty$ with P_i-probability 1 and $\int_0^\infty P(t, i, i)\, dt = \infty$.*

PROOF. *Claim (a).* Let $\tau_0 = 0$ and let τ_{n+1} be the least $t \geq \tau_n + 1$ with $X(t) = i$; if no such t exists, let $\tau_{n+1} = \infty$. Thus, $\tau_1 = \tau$. The strong Markov property $(MC, 9.41)$ implies

$$P_i\{\tau_{n+1} < \infty\} = f_i \cdot P_i\{\tau_n < \infty\}.$$

Indeed, τ_n is Markov. Let S be the post-τ_n process, with time domain retracted to the binary rationals. Clearly, $X(\tau_n) = i$ on $\{\tau_n < \infty\}$. And on Δ_q,

$$\{\tau_{n+1} < \infty\} = \{\tau_n < \infty\} \cap \{S \in [\tau_1 < \infty]\}.$$

Consequently, $P_i\{\tau_{n+1} < \infty\} = f_i^{n+1}$. Now

$$\{S_i \text{ is unbounded}\} \subset \{\tau_{n+1} < \infty\},$$

so S_i is bounded with P_i-probability 1. Next, let

$$L_n = \text{Lebesgue } \{S_i \cap [\tau_n, \tau_{n+1}]\}.$$

Since

$$\text{Lebesgue } \{S_i \cap [\tau_n + 1, \tau_{n+1}]\} = 0,$$

conclude $L_n \leq 1$. Let N be the greatest n with $\tau_n < \infty$. Thus,

$$\text{Lebesgue } S_i = \Sigma_{n=0}^{N-1} L_n \leq N.$$

But $\{N \geq n\} = \{\tau_n < \infty\}$, so $P_i\{N \geq n\} = f_i^n$ which sums, and N has finite P_i-expectation.

Claim (b) is argued similarly. Use strong Markov to check that the L_n are independent and identically distributed for $n = 0, 1, \ldots$. Indeed, L_0 is measurable on the pre-τ_1 sigma field. Let S be the post-τ_1 process, with time domain retracted to the binary rationals. On $\{\Delta_q \text{ and } \tau_1 < \infty\}$,

$$L_{n+1} = L_n \circ S \quad \text{for } n = 0, 1, \ldots.$$

Therefore, L_0 is P_1-independent of $(L_1, L_2, \ldots)$. And the P_i-distribution of $(L_1, L_2, \ldots)$ coincides with the P_i-distribution of $(L_0, L_1, \ldots)$. Use $(MC, 9.28)$ to check $P_i\{L_0 > 0\} > 0$. Conclude

$$P_i\{\Sigma_n L_n = \infty\} = 1. \qquad \qquad \bigstar$$

†For the definition, see page 304 of *MC*.

Call i *recurrent* or *transient* according as $f_i = 1$ or $f_i < 1$. Let

$$\{\text{hit } j\} = \{\omega : X(t, \omega) = j \text{ for some } t\}$$

$$= \{\omega : X(t, \omega) = j \text{ for some binary rational } t\}.$$

(23) Theorem. *Suppose $i \neq j$ and i is recurrent and $P_i\{\text{hit } j\} > 0$. Then $P_i\{\text{hit } j\} = P_j\{\text{hit } i\} = 1$ and j is recurrent.*

PROOF. Define τ_n as in (22), so $P_i\{\tau_n < \infty\} = 1$ for all n. There is a constant, finite N so large that

$$0 < P_i\{X(t) = j \text{ for some } t \leq \tau_N\}.$$

By strong Markov, the events

$$A_m = \{X(t) = j \text{ for some } t \text{ with } \tau_{mN} \leq t \leq \tau_{(m+1)N}\}$$

are independent and have common probability, for $m = 0, 1, \ldots$. Thus,

$$P_i\{\limsup A_m\} = 1.$$

Complete the proof as in (MC, 1.55).   ★

NOTE. $P_i\{\text{hit } j\} > 0$ iff $P(t, i, j) > 0$ for some t. By (2.8), either $P(t, i, j) = 0$ for all t or $P(t, i, j) > 0$ for all $t > 0$.

Say i *communicates* with j iff $P_i\{\text{hit } j\} > 0$ and $P_j\{\text{hit } i\} > 0$.

5. RESTRICTING THE RANGE

For the rest of this chapter, unless noted otherwise, let I be a finite or a countably infinite set in the discrete topology. Let $\bar{I} = I$ for finite I. Let $\bar{I} = I \cup \{\varphi\}$ be the one-point compactification of I for infinite I. Let P be a standard stochastic semigroup on I. As in (MC, Chap. 9), the process $\{X(t) : 0 \leq t < \infty\}$ on probability triple (Ω_m, P_i) is a Markov chain with stationary transitions P, starting state i, quasiregular sample functions, and metrically perfect level sets $S_j(\omega) = \{t : X(t, \omega) = j\}$ for all j in I. The relevant results are (9.16) and (9.28) of MC. For the rest of this chapter, unless noted otherwise, make the

(24) Assumption. Each i in I is recurrent, and communicates with each j in I.

This assumption eliminates large quantities of technical mess. The reasoning is, however, easily transferred to the general case, as indicated in Section 2.7.
    Let Ω_∞ be the set of $\omega \in \Omega_m$ such that

$$\text{Lebesgue } S_j(\omega) = \infty$$

for all j in I, and Lebesgue $S_\varphi(\omega) = 0$. Theorems (22), (23), and $(MC, 9.26)$ imply $P_i(\Omega_\infty) = 1$. Confine ω to Ω_∞ unless specified otherwise. It may be worth noting that for almost all but not all $\omega \in \Omega_\infty$, the set $S_j(\omega)$ is either nowhere dense or a union of intervals which do not accumulate, according as j is instantaneous or stable. The relevant results are (9.23) and (9.21) of MC. It may be worthwhile to look at (9.20) and (9.24) of MC again†. Let J be a variable finite subset of I. Let

$$S_J(\omega) = \{t : X(t, \omega) \in J\} = \cup_{j \in J} S_j(\omega),$$

so $S_{\{j\}} = S_j$. By quasiregularity, $S_J(\omega)$ is closed from the right, and each of its points is a limit of the set from the right. Consequently, $[0, \infty) \backslash S_J(\omega)$ is a union of intervals $[a, b)$ whose closures $[a, b]$ are disjoint. Let

$$\mu_J(t, \omega) = \text{Lebesgue } \{[0, t] \cap S_J(\omega)\}.$$

Thus, $\mu_J(\cdot, \omega)$ is continuous, nondecreasing, and flat off $S_J(\omega)$. Moreover, $\mu_J(\cdot, \omega)$ is strictly increasing on $S_J(\omega)$. Finally, $\mu_J(0, \omega) = 0$ and $\mu_J(\infty, \omega) = \lim_{t \to \infty} \mu_J(t, \omega) = \infty$. Let

$$\gamma_J(t, \omega) \text{ be the greatest } s \text{ with } \mu_J(s, \omega) = t.$$

Thus, $\gamma_J(\cdot, \omega)$ is continuous from the right, strictly increasing, and its range is exactly $S_J(\omega)$. Moreover, $\gamma_J(0, \omega)$ is the least t with $X(t, \omega) \in J$, and $\gamma_J(\infty, \omega) = \infty$. Finally, $\gamma_J(t, \omega) \geq t$ because $\mu_J(t, \omega) \leq t$. Most of this is worth preserving.

(25) Lemma.

    (a) $\mu_J(t) \leq t$.

    (b) $\gamma_J(t) \geq t$.

    (c) μ_J is continuous and nondecreasing, and is a strictly increasing map of S_J onto the time domain $[0, \infty)$ of X_J. In particular, μ_J is $1-1$ on S_J.

    (d) γ_J is a right continuous and strictly increasing map of the time domain $[0, \infty)$ of X_J onto S_J. In particular, γ_J is $1-1$.

    (e) $\mu_J[\gamma_J(t)] = t$.

    (f) $\gamma_J[\mu_J(t)] = t$ for $t \in S_J$.

    (g) $t - \mu_J(t) \leq \gamma_J(t) - t$.

PROOF.   The only new point is (g). Let

$$\theta(t) = t - \mu_J(t) = \text{Lebesgue } \{s : 0 \leq s \leq t \text{ and } X(s) \notin J\}.$$

Then θ is nondecreasing. So

$$\theta(t) \leq \theta[\gamma_J(t)] \qquad \text{by (b)}$$
$$= \gamma_J(t) - t \qquad \text{by (e).} \qquad \bigstar$$

†for a description of $S_j(\omega)$.

(26) Definition. *The restriction X_J of X to J is the process*

$$X_J(t) = X[\gamma_J(t)]:$$

namely,

$$X_J(t, \omega) = X[\gamma_J(t, \omega), \omega].$$

This is the main object of study for this chapter. Informally, X_J is X watched only when in J. The remaining times are simply erased from the time axis, which is then collapsed to form a continuous line again, as in Figure 2. Thus, $\mu_J(t)$ is the time on the X_J-scale corresponding to time t

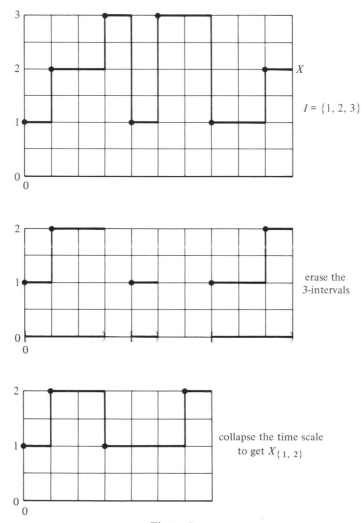

Figure 2.

on the X-scale. And $\gamma_J(t)$ is the greatest time on the X-scale corresponding to time t on the X_J-scale. As J increases to I, the process X_J approximates X more and more closely. This will be explored in the present chapter. Theorems and examples can be obtained for the infinite case by passing to the limit of the finite case. This will be explored in Chapter 2–3.

In the rest of this section, I will collect a number of technical facts which will be useful later. The main one is (32), and the most interesting one is (35).

(27) Lemma. *X_J has right continuous J-valued step functions for sample functions.*

PROOF.   Easy.   ★

For (28), let $i \in J$ and let ϕ be the first time X hits $J \setminus \{i\}$. Let τ_J be the first holding time in X_J.

(28) Lemma. *If ω is confined to $\{X(0) = i\}$, then*

$$\phi = \gamma_J(\tau_J) \quad and \quad \tau_J = \mu_J(\phi).$$

PROOF.   Use (25d) for the first claim, and (25e) for the second.

For (29), let $\{C_n\}$ be the complementary intervals of S_J. Plainly, μ_J is constant on C_n; let c_n be its value. Let f_n be the length of C_n. Let

$$F_J(t) = \Sigma \{ f_n : c_n \leqq t \}.$$

(29) Lemma. *$\gamma_J(t) = t + F_J(t)$, and F_J is discrete.*

PROOF.   Let $\gamma_J(t) = t + s$. Now $\mu_J(t + s) = t$ and $t + s \in S_J$. By the additivity of Lebesgue measure,

$$t + s = \mu_J(t + s) + \Sigma \{ f_n : C_n \subset [0, t + s] \}.$$

But $C_n \subset [0, t + s]$ implies $c_n \leqq \mu_J(t + s)$, because μ_J is nondecreasing; and $c_n \leqq \mu_J(t + s)$ implies $C_n \subset [0, t + s]$, because μ_J is strictly increasing on S_J, and $t + s \in S_J$. Thus, $s = F_I(t)$.   ★

To state (30), let G be a right continuous and nondecreasing function. Then G is the distribution function of a measure, say $|G|$. Let B be a Borel subset of the line.

(30) Lemma. *The Lebesgue measure of the G-image of B is the mass which the continuous part of $|G|$ assigns to B.*

PROOF.   Check it for intervals, and extend.   ★

(31) Corollary. *γ_J preserves Lebesgue measure. Namely, for Borel subsets B of $[0, \infty)$, the Lebesgue measure of B coincides with the Lebesgue measure of the γ_J-image of B.*

PROOF. Use (29) and (30): remember $|F_J|$ has no continuous part.   ★

Let K be a finite subset of I which includes J. Let

$$\mu_{J,K}(t) = \text{Lebesgue } \{s: 0 \leq s \leq t \text{ and } X_K(s) \in J\}$$

and let

$$\gamma_{J,K}(t) \text{ be the greatest } s \text{ with } \mu_{J,K}(s) = t.$$

Thus, $\mu_{J,K}(t)$ is the X_J-time corresponding to X_K-time t, and $\gamma_{J,K}(t)$ is the greatest X_K-time corresponding to X_J-time t.

(32) Corollary.   **(a)** $\mu_{J,K} = \mu_J(\gamma_K)$;

     (b) $\mu_{J,K}(\mu_K) = \mu_J$;

     (c) $\gamma_{J,K} = \mu_K(\gamma_J)$;

     (d) $\gamma_K(\gamma_{J,K}) = \gamma_J$.

PROOF. Claim (a) follows from (31). Indeed, let

$$A = \{s: 0 \leq s \leq t \text{ and } X[\gamma_K(s)] \in J\}$$

$$B = \{u: 0 \leq u \leq \gamma_K(t) \text{ and } X(u) \in J\}.$$

Then (a) is equivalent to the relation Lebesgue $A =$ Lebesgue B. I say $B = \gamma_K A$. Clearly, $\gamma_K A \subset B$. To argue the opposite inclusion, let $u \in B$. Then $X(u) \in J \subset K$, so $u = \gamma_K(s)$ for some s, because γ_K maps onto S_K. And $u \leq \gamma_K(t)$, so $s \leq t$, because γ_K is strictly increasing. Thus $u \in \gamma_K A$, and $B = \gamma_K A$. So (31) forces Lebesgue $A =$ Lebesgue B.

Now (a) implies

$$\mu_{J,K} \circ \mu_K = \mu_J \circ \gamma_K \circ \mu_K = \mu_J,$$

at least on S_K. But $\mu_{J,K} \circ \mu_K$ and μ_J are both continuous, and constant off S_K. This proves (b). Now (b) implies

$$\mu_{J,K} \circ \mu_K \circ \gamma_J = \mu_J \circ \gamma_J,$$

which is the identity function. But $\mu_K \circ \gamma_J$ is right continuous, proving (c). Apply γ_K to both sides of (c). Now $\gamma_K \circ \mu_K$ is the identity on S_K, so the relation $\gamma_J \in S_J \subset S_K$ proves (d).   ★

To be pedantic about $\mu_J(\gamma_K)$: this composition is a function of pairs $t \in [0, \infty)$ and $\omega \in \Omega_\infty$, whose value at (t, ω) is

$$\mu_J[\gamma_K(t, \omega), \omega].$$

Perhaps the best way of visualizing $\gamma_{J,K}(t)$ is that indicated by (c): the time X spends in K up to and including the tth instant in J.

(33) Lemma.   **(a)** $\mu_K(t) \uparrow t$ as $K \uparrow I$ for all $t \geq 0$.

     (b) $\gamma_K(t) \downarrow t$ as $K \uparrow I$ for all $t \geq 0$.

PROOF.   The second claim follows from the first, which is easy.   ★

Let $f(t-) = \lim_{s \downarrow 0} f(t-s)$. Thus, $\gamma_J(t-)$ is the least time on the X-scale corresponding to time t on the X_J-scale.

(34) Corollary.   (a) $\gamma_{J,K}(t) \uparrow \gamma_J(t)$ as $K \uparrow I$ for all $t \geq 0$.
         (b) $\gamma_{J,K}(t-) \uparrow \gamma_J(t-)$ as $K \uparrow I$ for all $t > 0$.
         (c) $\gamma_{J,K}(t) - \gamma_{J,K}(t-) \uparrow \gamma_J(t) - \gamma_J(t-)$ as $K \uparrow I$ for all $t > 0$.

PROOF.   Use (32c) and (33a) for (a) and (b). For (c), improve (33a) to:

$$\mu_K[a, b] \uparrow b - a \quad \text{as } K \uparrow I \quad \text{for} \quad 0 \leq a \leq b < \infty.\quad\quad ★$$

By definition, $(X_K)_J = X_K(\gamma_{J,K})$.

(35) Proposition. $(X_K)_J = X_J$.

PROOF.   Use (32d).   ★

That is, erasing first the non-K times in X and then the non-J times produces the same result as just erasing the non-J times.

EXAMPLE.   Even if P is uniform, it is possible that

$$\int \gamma_{\{i\}}(t) \, dP_i = \infty \quad \text{for all } t > 0.$$

CONSTRUCTION.   Let I consist of $i = 1$ and pairs (n, m) with $n = 1, 2, \ldots$ and $m = 1, \ldots, n$. Let $0 < p_n < 1$, with $\Sigma_{n=1}^{\infty} p_n = 1$ and $\Sigma_{n=1}^{\infty} np_n = \infty$. Let

$$Q(1, 1) = -1 \text{ and } Q[1, (n, 1)] = p_n$$

$$Q[(n, m), (n, m)] = -1 \text{ and } Q[(n, m), (n, m+1)] = 1 \text{ for } m = 1, \ldots, n-1$$

$$Q[(n, n), (n, n)] = -1 \text{ and } Q[(n, n), 1] = 1.$$

Let all other entries in Q vanish. By $(MC, 5.29)$, there is a uniform stochastic semigroup P on I with $P'(0) = Q$. Let θ be the first holding time in 1 for X, and let T be the length of the gap between the first and second 1-intervals in X. Use $(MC, 5.48)$ to see that with respect to P_1: the variables θ and T are independent; θ is exponential with parameter 1; and

$$\int T \, dP_1 = \Sigma \, np_n = \infty.$$

But

$$\gamma_{\{1\}}(t) \geq T \quad \text{on} \quad \{\theta > t\},$$

so

$$\int \gamma_{\{1\}}(t) \, dP_1 \geqq \int_{\{\theta > t\}} T \, dP_1$$

$$= P_1\{\theta > t\} \cdot \int T \, dP_1$$

$$= \infty. \qquad \qquad \bigstar$$

6. THE MARKOV PROPERTY

Continue in the setting of Section 5. Let J be a finite subset of I. Let $i \in J$.

(36) Lemma. X_J *has J-valued right continuous sample functions, in which each $j \in J$ is visited at a set of times of infinite Lebesgue measure.*

PROOF.   Use (27) and (32a). $\bigstar$

(37) Theorem. X_J *is Markov with stationary standard transitions, say P_J, on J and starting state i, relative to P_i.*

PROOF.   This follows from the strong Markov property $(MC, 9.41)$. Fix $s > 0$. Abbreviate $\sigma(\omega) = \gamma_J(s, \omega)$. As in Section 9.4 of MC, let $\mathscr{F}(t)$ be the σ-field spanned by $X(u)$ for $u \leq t$. Recall the meaning of Markov times from Section 9.4 of MC. I say that

$$(38) \qquad\qquad \sigma = \gamma_J(s) \quad \text{is a Markov time};$$

indeed, $\gamma_J(s) < t$ iff $s < \mu_J(t)$, and $\mu_J(t)$ is $\mathscr{F}(t)$-measurable. Let W be the set of $\omega \in \Omega_\infty$ such that $X[\sigma(\omega) + \cdot, \omega]$ is quasiregular. Let $S(\omega)$ be $X[\sigma(\omega) + \cdot, \omega]$ retracted to R, the binary rationals in $[0, \infty)$. Plainly, $\omega \in W$ entails $S(\omega) \in \Omega_\infty$. As $(MC, 9.41a)$ implies, $P_i(W) = 1$. For $\omega \in W$ and all $t \geqq 0$,

$$X[\sigma(\omega) + t, \omega] = X[t, S(\omega)];$$

consequently

$$(39) \qquad \mu_J[\sigma(\omega) + t, \omega] = \mu_J[\sigma(\omega), \omega] + \mu_J[t, S(\omega)]$$

$$= s + \mu_J[t, S(\omega)].$$

For $\omega \in W$ and all $t \geqq 0$, I assert

$$(40) \qquad\qquad \gamma_J(s + t, \omega) = \gamma_J(s, \omega) + \gamma_J[t, S(\omega)].$$

Indeed, let the left side of (40) be $\sigma(\omega) + \xi$, where $\xi = \xi(\omega) \geq 0$ by (25d). Then ξ is the greatest solution of

$$s + t = \mu_J[\sigma(\omega) + \xi, \omega]$$
$$= \mu_J[\sigma(\omega), \omega] + \mu_J[\xi, S(\omega)] \quad \text{by (39)}$$
$$= s + \mu_J[\xi, S(\omega)].$$

So $\xi = \gamma_J[t, S(\omega)]$, proving (40). In particular, for $\omega \in W$ and all $t \geq 0$,

(41) $$X_J(s + t, \omega) = X_J[t, S(\omega)].$$

Let $\mathcal{F}_J(t)$ be the σ-field spanned by $X_J(u)$ for $0 \leq u \leq t$. Recall the meaning of $\mathcal{F}(\sigma+)$ from Section 9.4 of MC. I claim

(42) $$\mathcal{F}_J(s) \subset \mathcal{F}(\sigma+).$$

I have to prove $B \cap \{\sigma < t\} \in \mathcal{F}(t)$ for all $B \in \mathcal{F}_J(s)$ and all t; for conventional reasons, I only have to do it for the generating B's, of the form $\{X_J(u) = j\}$. So let $u \leq s$ and $j \in J$. Then $\gamma_J(u) \leq \gamma_J(s) = \sigma$. Furthermore,

$$\{X_J(u) = j\} \cap \{\sigma < t\} = \{X[\gamma_J(u)] = j\} \cap \{\sigma < t\} = \bigcap_{n=1}^{\infty} A_n,$$

where A_n is the event that $\sigma < t$ and there is a binary rational r with

$$r - \frac{1}{n} \leq \gamma_J(u) < r < t \quad \text{and} \quad X(r) = j.$$

If $r < t$, then

$$\{\gamma_J(u) < r\} = \{\mu_J(r) > u\} \in \mathcal{F}(r) \subset \mathcal{F}(t).$$

Clearly, $X(r)$ is $\mathcal{F}(t)$-measurable for $r < t$. And $\{\sigma < t\} \in \mathcal{F}(t)$ by (38). So $A_n \in \mathcal{F}(t)$. This completes the proof of (42).

Let $P_J(t, i, j) = P_i\{X_J(t) = j\}$, for i and j in J. I will argue that X_J is Markov with transitions P_J. Fix nonnegative s and t. Fix i, j, k in J. Fix $A \in \mathcal{F}(s)$ with $A \subset \{X_J(s) = j\}$. Let $\sigma = \gamma_J(s)$. Then σ is Markov by (38), and $A \in \mathcal{F}(\sigma+)$ by (42), and $X(\sigma) = j$ on A. Remember $P_i(W) = 1$. Use strong Markov:

$$P_i\{A \text{ and } X_J(s + t, \cdot) = k\} = P_i\{A \text{ and } W \text{ and } X_J(s + t, \cdot) = k\}$$
$$= P_i\{A \text{ and } W \text{ and } X_J(t, S) = k\} \quad \text{by (41)}$$
$$= P_i\{A \text{ and } S \in [X_J(t, \cdot) = k]\}$$
$$= P_i\{A\} \cdot P_j\{X_J(t, \cdot) = k\} \qquad \text{by } (MC, 9.41c)$$
$$= P_i\{A\} \cdot P_J(t, j, k).$$

Now P_J is a stochastic semigroup, and X_J is Markov with transitions P_J starting from i relative to P_i, by $(MC, 5.4)$. Finally, P_J is standard by (36).   ★

(43) Definition.   *Call P_J the* restriction *of P to J, and $Q_J = P'_J(0)$ the restriction of Q to J.*

WARNING.   There is no easy way to compute P_J from P.

Of course, $Q = P'(0)$ does not determine Q_J. The existence of the derivatives in question was proved in $(MC, 5.21)$. Relative to P_J, each state in J is recurrent and communicates with each other state in J, by (36).

(44) Theorem.   *Fix $i \neq j$ in J. Relative to P_i, the time X spends in i until first hitting $J \setminus \{i\}$ is exponentially distributed with parameter $q_J(i) = -Q_J(i, i)$. Again relative to P_i, the probability that X hits j before $J \setminus \{i, j\}$ is $\Gamma_J(i, j) = Q_J(i, j)/q_J(i)$.*

This result is due to (Lévy, 1952).

PROOF.   Let τ_J be the first holding time in X_J, and let ϕ be the time at which X first hits $J \setminus \{i\}$. Confine ω to $\{X(0) = i\}$. Then (28) makes

$$\phi = \gamma_J(\tau_J);$$

so

$$\mu_{\{i\}}(\phi) = \mu_{\{i\}}[\gamma_J(\tau_J)]$$

$$= \mu_{\{i\}, J}(\tau_J) \quad \text{by (32a)}$$

$$= \tau_J \qquad \text{by thinking.}$$

Also,

$$X(\phi) = X[\gamma_J(\tau_J)] = X_J(\tau_J).$$

The P_i-distribution of τ_J and $X_J(\tau_J)$ can be obtained by using $(MC, 5.48)$ on X_J. Legitimacy is provided by (37).   ★

(45) Lemma.   *Let $J \subset K$ be finite subsets of I. Then*

$$(P_K)_J = P_J \quad and \quad (Q_K)_J = Q_J.$$

PROOF.   Use (37) and (35).   ★

7.   THE CONVERGENCE OF X_J TO X

Let J be a variable finite subset of I.

(46) Lemma.   *For j in J and $\delta > 0$,*

$$\int_0^\delta P_J(t, i, j)\, dt \geqq \int_0^\delta P(t, i, j)\, dt.$$

PROOF. $\int_0^\delta P_i\{X_J(t) = j\}\, dt = \int_{\Omega_\infty} \mu_{\{j\}, J}(\delta)\, dP_i$ by Fubini

$$= \int_{\Omega_\infty} \mu_{\{j\}}[\gamma_J(\delta)]\, dP_i \quad \text{by (32a)}$$

$$\geq \int_{\Omega_\infty} \mu_{\{j\}}(\delta)\, dP_i \qquad \text{by (25b)}$$

$$= \int_0^\delta P_i\{X(t) = j\}\, dt \quad \text{by Fubini.} \qquad \bigstar$$

(47) Proposition.   *The P_J are equicontinuous. Namely, for i in I and $\varepsilon > 0$, there is a $\delta = \delta(i, \varepsilon) > 0$, such that $P_J(t, i, i) \geq 1 - \varepsilon$ for $0 \leq t \leq \delta$ and all J containing i.*

PROOF.   Fix $\varepsilon > 0$. Because P is standard, there is a positive δ with

$$\frac{1}{\delta} \int_0^\delta P(t, i, i)\, dt \geq 1 - \tfrac{1}{2}\varepsilon.$$

Let $i \in J$. Use (46):

$$\frac{1}{\delta} \int_0^\delta P_J(t, i, i)\, dt \geq 1 - \tfrac{1}{2}\varepsilon.$$

Use (4) on P_J, with time rescaled by δ:

$$P_J(t, i, i) \geq 1 - \varepsilon \quad \text{for } 0 \leq t \leq \delta. \qquad \bigstar$$

NOTE. $(MC, 5.9)$ now gives an estimate for the modulus of continuity of $P_J(t, i, j)$.

(48) Theorem.   *For each $t > 0$, as J increases to I, the generalized sequence $X_J(t)$ converges to $X(t)$ in P_i-probability.*

PROOF.   Use the notation and result of the Markov property $(MC, 9.31)$. Confine ω to $W_t \cap \Omega_\infty$, which has P_i-probability 1. Then

(49a) $\mu_J(t + s, \omega) = \mu_J(t, \omega) + \mu_J(s, T_t\omega)$ for all $s \geq 0$.

Put $\varepsilon_J(t, \omega) = t - \mu_J(t, \omega)$, and deduce

(49b) $\gamma_J(t, \omega) = t + \gamma_J[\varepsilon_J(t, \omega), T_t\omega]$;

because $\gamma_J(t, \omega)$ is t plus the largest s with $\mu_J(t + s, \omega) = t$. In particular,

(50) $X_J(t, \omega) = X_J[\varepsilon_J(t, \omega), T_t\omega]$.

Write E_i for P_i-expectation. I claim

(51) $P_i\{X_J[\varepsilon_J(t), T_t] = X(t)\} = E_i\{P_J[\varepsilon_J(t), X(t), X(t)] \cdot 1_J[X(t)]\}$.

To prove (51), define F on $\Omega_\infty \times \Omega_\infty$ as follows:

$$F(\omega, \omega') = 1 \quad \text{when } X_J[\varepsilon_J(t, \omega), \omega'] = X(t, \omega)$$

$$= 0 \quad \text{elsewhere.}$$

Let $\mathscr{F}$ be the product σ-field in Ω relativized to Ω_∞. Both $\varepsilon_J(t)$ and $X(t)$ are $\mathscr{F}(t)$-measurable. So F is $\mathscr{F}(t) \times \mathscr{F}$-measurable. Recognize $F(\cdot, T_t)$ as the indicator of

$$X_J[\varepsilon_J(t), T_t] = X(t)$$

and use $(MC, 9.31g)$. The left side of (51) is

$$\int F(\omega, T_t\omega)\, P_i(d\omega) = \int F^*(\omega)\, P_i(d\omega).$$

To recognize F^*, fix ω; abbreviate $j = X(t, \omega)$ and $s = \varepsilon_J(t, \omega)$. Then

$$F^*(\omega) = \int F(\omega, \omega')\, P_{X(t,\omega)}(d\omega)$$

$$= P_j\{\omega' : X_J(s, \omega') = j\}$$

$$= P_J(s, j, j) \cdot 1_J(j).$$

This proves (51).

In view of (50–51),

(52) $$P_i\{X_J(t) = X(t)\} = E_i\{P_J[\varepsilon_J(t), X(t), X(t)] \cdot 1_J[X(t)]\}.$$

Fix $\varepsilon > 0$. Choose a finite subset F of I so large that

$$P_i\{X(t) \in F\} > 1 - \varepsilon.$$

Use (47) to choose a positive δ so small that

$$P_J(u, j, j) > 1 - \varepsilon$$

for all $j \in F$ and $0 \leq u \leq \delta$ and $J \supset F$. Now $\varepsilon_J(t) \downarrow 0$ as $J \uparrow I$ by (33a). Find a finite set G with $F \subset G \subset I$, such that for all $J \supset G$,

$$P_i\{\varepsilon_J(t) \leq \delta\} > 1 - \varepsilon.$$

For $J \supset G$, relation (52) implies that $P_i\{X_J(t) = X(t)\}$ is at least

$$P_i\{X(t) \in F \text{ and } \varepsilon_J(t) \leq \delta\} \cdot \min\{P_J(u, j, j) : j \in F \text{ and } u \leq \delta\},$$

which exceeds

$$(1 - 2\varepsilon)(1 - \varepsilon). \qquad\qquad \bigstar$$

(53) Corollary. (a) $\lim_{J \uparrow I} P_J(t, i, j) = P(t, i, j)$ *uniformly on compact t-sets.*
   (b) $\{Q_J : finite\ J \subset I\}$ *determines P.*
   (c) $P_i\{q\text{-}\lim_{J \uparrow I} X_J(t) = X(t)\} = 1$ *for each* $t \geq 0$.

PROOF. *Claim (a).* Use (48) to get pointwise convergence and (47) to get the uniformity.
   Claim (b). Use (37) and $(MC, 5.29)$ to see that Q_J determines P_J. Then use (a).
   Claim (c). Relation (33b) and quasiregularity imply

$$X^*(t) = q\text{-}\lim_{J \uparrow I} X_J(t)$$

exists for all $t \geq 0$. But $X_J(t)$ converges to $X(t)$ in P_i-probability by (48). This forces

$$P_i\{X^*(t) = X(t)\} = 1. \qquad \bigstar$$

At one time I thought $X^*(t) = X(t)$ for all $t \geq 0$, perhaps after discarding a null set. But the proof disappeared. I doubt that (53c) holds for ordinary limits, but have no counter-example.

A continuity result

The continuity fact (54) will be used in Section 3.6. On a first reading, skip the proof. Fix $J \subset I$ and Q_J, so P_J is fixed. Let I and P vary along a sequence Σ. Unspecified limits are along Σ.

(54) Theorem. *Let* $i \in J$ *and* $T > 0$. *Suppose* $P_i\{\gamma_J(T) < T + \varepsilon\} \to 1$ *for all* $\varepsilon > 0$. *Then* $P(s, j, k) \to P_J(s, j, k)$ *uniformly on compact s-sets, for all j and k in J.*

PROOF. I claim that for all positive t less than T, all positive ε, and all j in J,

(55a) $P_j\{\gamma_J(t) < t + \varepsilon\} \to 1.$

Indeed, let $s = T - t$; let $\sigma = \gamma_J(s)$; let W be the set of ω where the function $X[\sigma(\omega) + \cdot, \omega]$ is quasiregular; and let $S\omega$ be $X[\sigma(\omega) + \cdot, \omega]$ retracted to the nonnegative binary rationals. Then σ is Markov by (38) and $P_i(W) = 1$ by $(MC, 9.41a)$. By (40),

$$\gamma_J(T, \omega) = \gamma_J(s, \omega) + \gamma_J(t, S\omega) \quad \text{for } \omega \in W.$$

But $\gamma_J(s, \omega) \geq s$, so $\gamma_J(T, \omega) < T + \varepsilon$ implies $\gamma_J(t, S\omega) < t + \varepsilon$. Therefore,

$$P_i\{\gamma_J(t, S) < t + \varepsilon\} \to 1.$$

Remember that $X_J(s) = X[\gamma_J(s)] = X(\sigma)$. Use strong Markov $(MC, 9.41)$.

$$P_j\{\gamma_J(t) < t + \varepsilon\} = P_i\{\gamma_J(t, S) < t + \varepsilon | X_J(s) = j\} \to 1.$$

Check $P_J(s, i, j) > 0$: because J is finite, and i leads to j relative to P_J. If you have trouble, look at Sections 2.3–4. This completes the proof of (55a).

Now I claim

(55b)
$$\int_0^t P(s, j, k)\, ds \rightarrow \int_0^t P_J(s, j, k)\, ds$$

for all $t < T$ and all j, k in J. Indeed, by Fubini,

$$\int_0^t P_J(s, j, k)\, ds = \int_{\Omega_\infty} \mu_{\{k\}, J}(t)\, dP_j = \alpha + \beta,$$

where

$$\alpha = \int_{\{\gamma_J(t) < t + \varepsilon\}} \mu_{\{k\}, J}(t)\, dP_j$$

and

$$\beta = \int_{\{\gamma_J(t) \geqq t + \varepsilon\}} \mu_{\{k\}, J}(t)\, dP_j.$$

But

$$\mu_{\{k\}, J}(t) = \mu_{\{k\}}[\gamma_J(t)] \quad \text{by (32a)}$$
$$\leqq \mu_{\{k\}}(t + \varepsilon) \quad \text{on } \{\gamma_J(t) < t + \varepsilon\}.$$

So

$$\alpha \leqq \int_{\Omega_\infty} \mu_{\{k\}}(t + \varepsilon)\, dP_j = \int_0^{t + \varepsilon} P(s, j, k)\, ds.$$

And $\mu_{\{k\}, J}(t) \leqq t$, so $\beta \leqq t P_j\{\gamma_J(t) \geqq t + \varepsilon\}$. Therefore,

$$\int_0^t P_J(s, j, k)\, ds \leqq \int_0^{t + \varepsilon} P(s, j, k)\, ds + t P_j\{\gamma_J(t) \geqq t + \varepsilon\}.$$

By (46),

$$\int_0^t P_J(s, j, k)\, ds \geqq \int_0^t P(s, j, k)\, ds.$$

This proves (55b).

Specialize $k = j \in J$ in (55b) and use (4):

$P(\cdot, j, j)$ is equicontinuous, as (I, P) varies through Σ.

By $(MC, 5.9)$, for j and k in J,

$P(\cdot, j, k)$ is equicontinuous, as (I, P) varies through Σ.

Use Arzela–Ascoli (Dunford and Schwarz, 1958, page 266) and the diagonal argument (*MC*, Sec. 10.12): for any subsequence of Σ, there is an R and a sub-subsequence along which

(55c) $$P(s, j, k) \rightarrow R(s, j, k)$$

uniformly on compact s-sets, for all j and k in J. In principle, R depends on the sub-subsequence. Clearly, $R(\cdot, j, k)$ is nonnegative, continuous, and 1 or 0 at 0, according as $j = k$ or $j \neq k$. By (55b),

$$\int_0^t R(s, j, k)\, ds = \int_0^t P_J(s, j, k)\, ds \quad \text{for all } t < T.$$

Consequently,

(55d) $$R = P_J \quad \text{on } [0, T].$$

Clearly,

$$P(t + s, j, k) \geq \Sigma_{h \in J}\, P(t, j, h) P(s, h, k).$$

Use (55c):

(55e) $$R(t + s, j, k) \geq \Sigma_{h \in J}\, R(t, j, h) R(s, h, k).$$

Of course,

(55f) $$P_J(t + s, j, k) = \Sigma_{h \in J}\, P_J(t, j, h) P_J(s, h, k).$$

By (55d–f),

(55g) $$R(t, j, k) \geq P_J(t, j, k)$$

for all t. But $R(t)$ is substochastic by Fatou, so equality must hold in (55g). ★

8. THE DISTRIBUTION OF γ_J GIVEN X_J

Recall that

$$\mu_J(t) = \text{Lebesgue } \{t : 0 \leq s \leq t \text{ and } X(s) \in J\};$$

γ_J is the right continuous inverse of μ_J; and $X_J = X(\gamma_J)$. Let X_J visit states $\xi_{J,0}, \xi_{J,1}, \ldots$ with holding times $\tau_{J,0}, \tau_{J,1}, \ldots$, as in Figure 3. Fix J with at least two elements and fix i in J. Recall that $f(t-) = \lim_{s \downarrow 0} f(t - s)$.

(56) **Theorem.** *On a suitable probability triple, there is a random variable T and a process $\{G(t) : 0 \leq t < \infty\}$ with the following properties:*

(a) *The joint distribution of T and $G(T-)$ coincides with the joint P_i-distribution of $\tau_{J,0}$ and $\gamma_J(\tau_{J,0}-)$.*

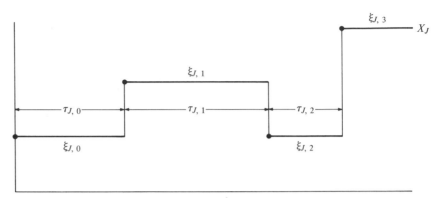

Figure 3.

(b)   *T and G are independent.*

(c)   *G has stationary, independent increments.*

(d) $G(0) = 0$.

(e) $\lim_{t \to 0} G(t)/t = 1$.

(f)   *The process* $\{G(t) - t : 0 \leq t < \infty\}$ *has right continuous and nondecreasing sample functions.*

It is known that if G satisfies (c-f), then G has no fixed points of discontinuity. Thus $G(T-) = G(T)$ a.e. However, $\gamma_J(\tau_{J,0}-)$ is the least time on the X-scale corresponding to time $\tau_{J,0}$ on the X_J-scale, and has a good chance of being really less than $\gamma_J(\tau_{J,0})$. Of course, G and T depend on i and J.

PROOF.   Let $0 < a < b < \infty$. Remember $i \in J$. I maintain that

(57)   with respect to P_i, given $\tau_{J,0} \geq b$, the process $\{\gamma_J(t) : 0 \leq t \leq a\}$ has stationary independent increments, whose distribution does not depend on b.

Indeed, let $0 = a_0 < a_1 < \cdots < a_n < a$. Let $x_1, \ldots, x_n$ be nonnegative real numbers. Let A be the event

$$\gamma_J(a_{m+1}) - \gamma_J(a_m) > x_{m+1} \quad \text{for } m = 0, \ldots, n - 1.$$

My job is to compute $P_i \{A \text{ and } \tau_{J,0} \geq b\}$. More exactly, I have to exhibit a function $f(\cdot, \cdot)$ such that

(58a) $P_i\{A \text{ and } \tau_{J,0} \geq b\} = [\Pi_{m=0}^{n-1} f(a_{m+1} - a_m, x_{m+1})] \cdot P_i\{\tau_{J,0} \geq b\}.$

Let

(58b) $B(a) = \{\tau_{J,0} > a\}.$

Then (58a) holds with

(58c) $f(a, x) = g(a, x) \exp [q_J(i)a]$,

where

(58d) $g(a, x) = P_i\{B(a) \text{ and } \gamma_J(a) > x\}$

and

$$\exp y = e^y.$$

From (37) and $(MC, 5.48)$,

(59) $P_i\{B(a)\} = \exp [-q_J(i)a]$.

So $1 - f(a, \cdot)$ is a distribution function.

I will prove this sharp form of (58) inductively. Let $\hat{A}$ be the event

$$\gamma_J(a_{m+1} - a_1) - \gamma_J(a_m - a_1) > x_{m+1} \quad \text{for } m = 1, \ldots, n-1.$$

So $\hat{A} = \Omega_\infty$ when $n = 1$. The main step in proving (58) is

(60) $P_i\{A \text{ and } \tau_{J,0} \geqq b\} = g(a_1, x_1) \cdot P_i\{\hat{A} \text{ and } \tau_{J,0} \geqq b - a_1\}$.

For then

$$P_i\{A \text{ and } \tau_{J,0} \geqq b\} = [\Pi_{m=0}^{n-1} g(a_{m+1} - a_m, x_{m+1})] \cdot P_i\{\tau_{J,0} \geqq b - a_n\}$$

$$= [\Pi_{m=0}^{n-1} f(a_{m+1} - a_m, x_{m+1})] \cdot P_i\{\tau_{J,0} \geqq b\},$$

using (59).

First, $b > a_1$ makes

(61) $\{\tau_{J,0} \geqq b\} \subset B(a_1) = \{\tau_{J,0} > a_1\}$.

I will now argue (60) from strong Markov. Abbreviate $\sigma = \gamma_J(a_1)$. Let W be the set of $\omega \in \Omega_\infty$ such that $X[\sigma(\omega) + \cdot, \omega]$ is quasiregular. For $\omega \in W$, let $S(\omega)$ be $X[\sigma(\omega) + \cdot, \omega]$ retracted to R, the set of binary rationals in $[0, \infty)$. Put $s = a_1$ in (38–42): so σ, W, and S agree. Then σ is a Markov time by (38). And $P_i(W) = 1$ by strong Markov $(MC, 9.41a)$. Next,

(62a) $B(a_1) \in \mathscr{F}_J(a_1) \subset \mathscr{F}(\sigma+)$

by (42). And $X(\sigma) = X_J(a_1)$, so

(62b) $X(\sigma) = i \quad \text{on } \{X(0) = i \text{ and } B(a_1)\}$.

Clearly,

(62c) $\{\gamma_J(a_1) > x_1\} \in \mathscr{F}(\sigma+)$.

Fix $\omega \in W$. If $\tau_{J,0}(\omega) > a_1$, then $\tau_{J,0}(\omega)$ is a_1 plus the first holding time in $X_J(a_1 + \cdot, \omega)$. By (41),

(62d) $$\tau_{J,0} = a_1 + \tau_{J,0} \circ S \quad \text{on } W \cap B(a_1).$$

Fix $\omega \in W$ and m. By (40),

$$\gamma_J(a_m, \omega) = \gamma_J(a_1, \omega) + \gamma_J(a_m - a_1, S(\omega)).$$

So

(62e) $\gamma_J(a_{m+1}) - \gamma_J(a_m) = \gamma_J(a_{m+1} - a_1) \circ S - \gamma_J(a_m - a_1) \circ S \quad$ on W.

Clearly,

(62f) $$\gamma_J(0) = 0 \quad \text{on } \{X(0) = i\}.$$

Combine (62d–f) to see that on $\{X(0) = i \text{ and } W\}$,

(63) $\{A \text{ and } \tau_{J,0} \geq b\} = \{B(a_1) \text{ and } \gamma_J(a_1) > x_1\} \cap S^{-1}\{\hat{A} \text{ and } \tau_{J,0} \geq b - a_1\}.$

Now use (62a–c) and strong Markov (MC, 9.41c) to get (60).

On a convenient probability triple, construct a process $G(r)$ for nonnegative rational r, and a random variable T, such that:

  T and G are independent;
  T is exponential with parameter $q_J(i)$;
  $G(0) = 0$;
  G has stationary, independent increments;
  Prob $\{G(r) > x\} = f(r, x)$, as defined in (58b–d)

Use (58), (59), and (62f): the distribution of

$$T \quad \text{and} \quad G(r) \text{ for nonnegative rational } r < T$$

coincides with the P_i-distribution of

$$\tau_{J,0} \quad \text{and} \quad \gamma_J(r) \text{ for nonnegative rational } r < \tau_{J,0}.$$

According to (29),

$$r \to \gamma_J(r) - r$$

is nondecreasing and continuous from the right. By discarding a null set,

$$r \to G(r) - r$$

is nondecreasing and continuous from the right: because T is unbounded. Extend G to the whole nonnegative line: let $G(t)$ be the limit of $G(r)$ as rational r decreases to t. Then G satisfies (56b–d, f). Because γ_J is continuous from the right,

(64)   The distribution of

$$T \quad \text{and} \quad G(t) \text{ for nonnegative real } t < T$$

coincides with the P_i-distribution of

$$\tau_{J,0} \quad \text{and} \quad \gamma_J(t) \text{ for nonnegative real } t < \tau_{J,0}.$$

This is sharper than (56a). Strict inequality occurs, because $\gamma_J(\tau_{J,0})$ cannot be computed from $\gamma_J(r)$ for rational $r < \tau_{J,0}$.
  Property (56e) is equivalent to

$$P_i\{\lim_{t \to 0} \gamma_J(t)/t = 1\} = 1.$$

This in turn is equivalent to $(MC, 9.32)$. Here is an independent proof of these assertions: its final form is due to Isaac Meilijson. To begin with, suppose F is nondecreasing. Then

$$\lim \sup_{h \downarrow 0} \frac{1}{h}[F(t + h) - F(t)] = \lim \sup_{n \uparrow \infty} n\left[F\left(t + \frac{1}{n}\right) - F(t)\right].$$

Similarly for lim inf. This solves a joint measurability problem that will appear in a minute. Let F^+ be the right derivative of F. Relation (29) and $(MC, 10.61)$ imply $\gamma_J^+(t) = 1$ for Lebesgue almost all t. Fubini produces a Lebesgue null set N such that

$$P_i\{\gamma_J^+(t) = 1\} = 1 \quad \text{for } t \notin N.$$

By (64),
$$\text{Prob}\{G^+(t) = 1\} = 1 \quad \text{for } t \notin N.$$

Now (56c–d) get (56e).   ★

(64*)   NOTE.   If $J = \{i\}$, then $\tau_{J,0} = \infty$. The argument shows that γ_J by itself has properties (56c–f). Property (c) for $J = \{i\}$ is due to (Lévy, 1952).

(65a)   Let $\alpha_{J,0} = \gamma_J(\tau_{J,0}-)$ and for $n = 1, 2, \ldots$ let

$$\alpha_{J,n} = \gamma_J(\tau_{J,0} + \cdots + \tau_{J,n}-) - \gamma_J(\tau_{J,0} + \cdots + \tau_{J,n-1}).$$

(65b)   For $n = 1, 2, \ldots$ let

$$\beta_{J,n} = \gamma_J(\tau_{J,0} + \cdots + \tau_{J,n-1}) - \gamma_J(\tau_{J,0} + \cdots + \tau_{J,n-1}-).$$

Thus, $\alpha_{J,n}$ is the time X spends interior to the nth interval of constancy for X_J. And $\beta_{J,n}$ is the time X spends at the nth discontinuity of X_J. The intervals are numbered from 0 and the discontinuities from 1. Theorem (56) and the next result give considerable information on the distribution of γ_J given X_J. This is enough for present applications, but a more complete description is in Section 11.

(66)   Theorem.   *Suppose $N = 0, 1, 2, \ldots.$ Given $\xi_{J,0}, \xi_{J,1}, \ldots, \xi_{J,N}$:*

(a)   *$(\tau_{J,0}, \alpha_{J,0}), \beta_{J,1}, (\tau_{J,1}, \alpha_{J,1}), \beta_{J,2}, \ldots, (\tau_{J,N}, \alpha_{J,N}), \beta_{J,N+1}$ are conditionally P_i-independent;*

(b)   *for $1 \leq n \leq N$, the conditional P_i-distribution of $(\tau_{J,n}, \alpha_{J,n})$ on $\{\xi_{J,n} = j\}$ coincides with the unconditional P_j-distribution of $(\tau_{J,0}, \alpha_{J,0})$;*

(c)   *for $1 \leq n \leq N$, the conditional P_i-distribution of $\beta_{J,n}$ on the set $\{\xi_{J,n-1} = j \text{ and } \xi_{J,n} = k\}$ coincides with the P_j-distribution of $\beta_{J,1}$ given $\{\xi_{J,1} = k\}$.*

PROOF.   Use (109) below.   ★

An alternative form of (66)

Using (MC, 10.10a and 16), this theorem can be restated in an equivalent and more constructive looking form. For $j \in I$, let $D(j, \cdot)$ be the P_j-distribution of $(\tau_{J,0}, \alpha_{J,0})$; so $D(j, \cdot)$ is a probability on the set of pairs of nonnegative numbers. For $j, k \in J$, let $D(j, k, \cdot)$ be the P_j-distribution of $\beta_{J,1}$ given $\{\xi_{J,1} = k\}$; so $D(j, k, \cdot)$ is a probability on the nonnegative numbers.

INTEGRATED FORM OF (66).   For all N, and $i_0 = i, i_1, \ldots, i_N$ in J and Borel A_n, B_n:

$$P_i\{\xi_{J,n} = i_n \text{ and } (\tau_{J,n}, \alpha_{J,n}) \in A_n \text{ for } n = 0, \ldots, N$$

$$\text{and } \beta_{J,n} \in B_n \quad \text{for } n = 1, \ldots, N\} = abc$$

where

$$a = \Pi_{n=0}^{N-1} \Gamma_j(i_n, i_{n+1})$$
$$b = \Pi_{n=0}^{N} D(i_n, A_n)$$
$$c = \Pi_{n=1}^{N} D(i_{n-1}, i_n, B_n).$$

9.   THE JOINT DISTRIBUTION OF $\{X_J\}$

Let $\mathscr{P} = (T_0, T_1, \ldots)$ be a point process in $(0, \infty]$. This means: the T_n are positive extended real-valued random variables; and $T_{n+1} > T_n$ when

$T_n < \infty$; and $T_{n+1} = \infty$ when $T_n = \infty$; and $T_n \to \infty$ as $n \to \infty$. If $0 \leq U < V \leq \infty$ are constants or even random variables, the *restriction* of $\mathscr{P}$ to (U, V) is the sequence of T_n's falling in (U, V); this sequence may be empty, and in general has a random number of terms. Define the process $X_{\mathscr{P}} = \{X_{\mathscr{P}}(t) : 0 \leq t < \infty\}$ as follows: $X_{\mathscr{P}}(t)$ is the least n with $T_n > t$. Of course, $X_{\mathscr{P}}(0) = 0$; the sample functions of $X_{\mathscr{P}}$ are right continuous and nondecreasing; and

(67)   the σ-field spanned by $X_{\mathscr{P}}$ coincides with the σ-field spanned by $\mathscr{P}$.

Suppose $\mathscr{P}_f$ is a point process for each $f \in F$. Then (67) implies

(68)   The $\mathscr{P}_f$ are independent iff the $X_{\mathscr{P}_f}$ are.

If $q = 0$, then $\mathscr{P}$ is a *Poisson process* with parameter q iff $T_n = \infty$ almost surely for all n. If $q > 0$, then $\mathscr{P}$ is a *Poisson process* with parameter q iff $T_0, T_1 - T_0, T_2 - T_1, \ldots$ are independent and exponential with parameter q. If $\mathscr{P}$ is Poisson with parameter q, then $(MC, 5.39)$ shows that

(69a) $X_{\mathscr{P}}$ has stationary, independent increments.

And an easy integration shows that

(69b) $$\mathrm{Prob}\,\{X_{\mathscr{P}}(t) = n\} = \frac{(qt)^n}{n!} e^{-qt},$$

the Poisson distribution. Since (69a–b) specify the distribution of $X_{\mathscr{P}}$, relation (67) shows:

(70)   If $X_{\mathscr{P}}$ has properties (69a–b), then $\mathscr{P}$ is Poisson with parameter q.

In the usual terminology, $X_{\mathscr{P}}$ is Poisson with parameter q iff it has properties (69a–b). A remarkable fact, which will not be used here, is:

$\mathscr{P}$ is Poisson (with unspecified parameter) if $X_{\mathscr{P}}$ satisfies (69a).

To state (71), let F be a finite set. Let p be a probability on F. Let $Z_0, Z_1, \ldots$ be a sequence of independent F-valued random variables, with common distribution p. Suppose $\mathscr{P}$ is a Poisson process with parameter q, independent of the Z's. Let $N(0, f) < N(1, f) < \cdots$ be the indices n with $Z_n = f$. If there are m or fewer such indices, let $N(m, f) = \infty$. Let $T_\infty = \infty$. Let

$$\mathscr{P}_f = (T_{N(0,f)}, T_{N(1,f)}, \ldots).$$

That is, $\mathscr{P}_f$ consists of those T_n with $Z_n = f$.

(71) Lemma.   *The process $\mathscr{P}_f$ is Poisson with parameter $qp(f)$. And the processes $\mathscr{P}_f$ are independent as f varies over F.*

PROOF.   The program is to use (70) for the Poissonity of $\mathscr{P}_f$, and (68) for the independence. Let $X(t) = X_{\mathscr{P}}(t)$ and $X_f(t) = X_{\mathscr{P}_f}(t)$ for $f \in F$. Let $M = 1, 2, \ldots$. Let $0 = t_0 < t_1 < \cdots < t_M$. Let $n_f(1) \leq n_f(2) < \cdots \leq n_f(M)$ be nonnegative integers, for $f \in F$. Let

$$A = \{X_f(t_m) = n_f(m) \text{ for } f \in F \text{ and } m = 1, \ldots, M\}.$$

Put $n(0) = 0$ and $n(m) = \Sigma_f \, n_f(m)$. Then $A = \bigcap_{m=1}^{M} (A_m \cap B_m)$, where

$$A_m = \{X(t_m) - X(t_{m-1}) = n(m) - n(m-1)\}$$

and B_m is the event that for all $f \in F$, the number of indices v such that $n(m-1) \leq v \leq n(m) - 1$ and $Z_v = f$ is $n_f(m) - n_f(m-1)$. Use (69a) and the independence of $Z_1, Z_2, \ldots, \mathscr{P}$:

$$\text{Prob } A = \Pi_{m=1}^{M} (\text{Prob } A_m) \cdot (\text{Prob } B_m).$$

Put $d(f, m) = n_f(m) - n_f(m-1)$ and $d(m) = \Sigma_f \, d(f) = n(m) - n(m-1)$ and $s(m) = t_m - t_{m-1}$. Use (69b) and the fact that $Z_1, Z_2, \ldots$ are independent with common distribution p:

$$\text{Prob } A_m = \frac{[qs(m)]^{d(m)}}{d(m)!} e^{-qs(m)}$$

and

$$\text{Prob } B_m = d(m)! \, \Pi_f \, \frac{p(f)^{d(f,m)}}{d(f,m)!}.$$

Consequently:

$$(\text{Prob } A_m) \cdot (\text{Prob } B_m) = \Pi_f \, \frac{[qp(f)s(m)]^{d(f,m)}}{d(f,m)!} e^{-qp(f)s(m)};$$

and

$$\text{Prob } A = \Pi_{f \in F} \, \Pi_{m=1}^{M} \, \frac{[qp(f)s(m)]^{d(f,m)}}{d(f,m)!} e^{-qp(f)s(m)}.$$

The last equality makes the X_f independent and Poisson with parameter $qp(f)$. Relations (68) and (70) clinch the argument.   ★

Return to the setting of Section 5. To simplify the writing, for the rest of this section let $I = \{1, 2, \ldots\}$ or $\{1, 2, \ldots, M\}$ and $I_n = \{1, 2, \ldots, n\}$. Write X_n, P_n, and Q_n for X_{I_n}, P_{I_n}, and Q_{I_n}. Let $q_n(i) = -Q_n(i, i)$ for $i \in I_n$. Let

(72a) $\Gamma_n(i, j) = Q_n(i, j)/q_n(i)$ for $i \neq j$ in I_n

                         $= 0$ for $i = j$ in I_n.

Let

(72b) $$\pi_n(i) = 1 - \Gamma_n(i, n)\Gamma_n(n, i).$$

If $n \geq 2$, assumption (24) implies that Γ_n is a stochastic matrix and $q_n > 0$. And $\pi_n > 0$ for $n \geq 3$.

WARNING. The symbol P_n now has two meanings: the semigroup P restricted to I_n; and the probability on the space of sample functions making the coordinate process Markov with transitions P and starting state n. The meaning is always clear from the context, but the probability will now be written in boldface: $\mathbf{P_n}$.

I can now state (73); the proof is deferred.

(73) Lemma. **(a)** $q_n(i) = \pi_{n+1}(i)q_{n+1}(i)$ *for i in* I_n.
For $i \neq j$ *in* I_n:
  (b) $\pi_{n+1}(i)\Gamma_n(i, j) = \Gamma_{n+1}(i, j) + \Gamma_{n+1}(i, n + 1)\Gamma_{n+1}(n + 1, j);$
  (c) $Q_n(i, j) = Q_{n+1}(i, j) + Q_{n+1}(i, n + 1)\Gamma_{n+1}(n + 1, j).$

To state (74–76), and for later use in this section, let X_n visit the states $\zeta_{n,0}, \zeta_{n,1}, \dots$ with holding times $\tau_{n,0}, \tau_{n,1}, \dots$, as in Figure 4. As (37) implies, the process X_{n+1} is Markov with stationary transitions P_{n+1} and starting state i, relative to $\mathbf{P_i}$. Now use $(MC, 5.48)$ to get:

(74) $\mathbf{P_i}\{\xi_{n+1,v} = i_v$ and $\tau_{n+1,v} > t_v$ for $v = 0, \dots, m\} = pe^{-t},$ where
   $i_0 = i, i_1, \dots, i_m$ are in I_{n+1}
   $t_0, t_1, \dots, t_m$ are nonnegative numbers
   $p = \Pi_{v=0}^{m-1} \Gamma_{n+1}(i_v, i_{v+1})$
   $t = \Sigma_{v=0}^{m} q_{n+1}(i_v)t_v.$

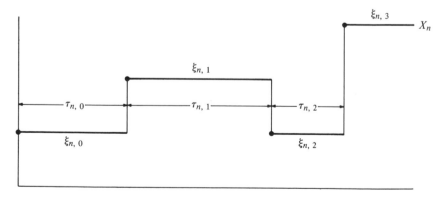

Figure 4.

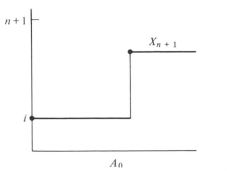

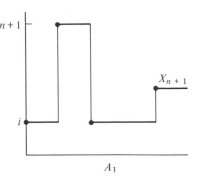

Figure 5.

To state (75–78), define subsets A_m and B_m of $\{X(0) = i\}$ as in Figures 5 and 6. Informally, A_m is the event that ξ_{n+1} executes m cycles $i \to n + 1 \to i$, and then exits from i to a state which is neither i nor $n + 1$. Formally,

$$A_m = \{\xi_{n+1,0} = \xi_{n+1,2} = \cdots = \xi_{n+1,2m} = i, \text{ and}$$

$$\xi_{n+1,1} = \xi_{n+1,3} = \cdots = \xi_{n+1,2m-1} = n + 1, \text{ but}$$

$$\xi_{n+1,2m+1} \text{ is neither } i \text{ nor } n + 1\}.$$

Informally, B_m is the event that ξ_{n+1} executes m cycles $i \to n + 1 \to i$, then moves to $n + 1$, and then exits from $n + 1$ to a state which is neither i nor $n + 1$. Formally,

$$B_m = \{\xi_{n+1,0} = \xi_{n+1,2} = \cdots = \xi_{n+1,2m} = i, \text{ and}$$

$$\xi_{n+1,1} = \xi_{n+1,3} = \cdots = \xi_{n+1,2m+1} = n + 1, \text{ but}$$

$$\xi_{n+1,2m+2} \text{ is neither } i \text{ nor } n + 1\}.$$

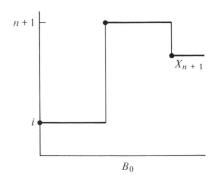

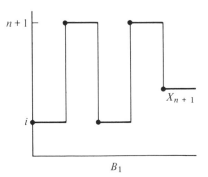

Figure 6.

Let $A = \bigcup_{m=0}^{\infty} A_m$ and $B = \bigcup_{m=0}^{\infty} B_m$. Of course, $A \cup B = \{X(0) = i\}$. Define the random variable M as follows: $M = m$ on $A_m \cup B_m$. As (35) implies:

(75) $\qquad \tau_{n,0} = \tau_{n+1,0} + \tau_{n+1,2} + \cdots + \tau_{n+1,2m}$ on $\{M = m\}$;

(76) $\qquad \xi_{n,1} = \xi_{n+1,2m+1}$ on A_m and $\xi_{n,1} = \xi_{n+1,2m+2}$ on B_m.

Furthermore, on $\{M = m\}$ define

$$\theta_0 = \tau_{n+1,0}$$
$$\theta_1 = \tau_{n+1,0} + \tau_{n+1,2}$$
$$\vdots$$
$$\theta_{m-1} = \tau_{n+1,0} + \tau_{n+1,2} + \cdots + \tau_{n+1,2m-2}.$$

If $m = 0$, there are no θ's. Thus: $\theta_0, \theta_1, \ldots$ are the times on the X_n-scale corresponding to $(n + 1)$-intervals of X_{n+1} interior to the initial interval of constancy for X_n.

(77) Lemma. *Fix i in I_n. On a convenient probability triple, construct a Poisson process $\mathscr{P}$ with parameter $[1 - \pi_{n+1}(i)]q_{n+1}(i)$. Construct a random variable τ on the same triple, which is exponentially distributed with parameter $\pi_{n+1}(i)q_{n+1}(i)$ and is independent of $\mathscr{P}$. Consider the joint distribution of τ and the restriction of $\mathscr{P}$ to $(0, \tau)$. This coincides with the joint $\mathbf{P}_i$-distribution of $\tau_{n,0}$ and $(\theta_0, \theta_1, \ldots)$.*

To state (78), let $n \geq 2$ and $i \neq j$ in I_n. Let

$$\hat{A}_m = \{A_m \text{ and } \xi_{n+1,2m+1} = j\} \quad \text{and} \quad \hat{B}_m = \{B_m \text{ and } \xi_{n+1,2m+2} = j\}.$$

Let $\hat{A} = \bigcup_{m=0}^{\infty} \hat{A}_m$ and $\hat{B} = \bigcup_{m=0}^{\infty} \hat{B}_m$. Confine ω to $\{X(0) = i\}$. Then $\hat{A} \cup \hat{B}$ is the event that $\xi_{n,1} = j$. And $\hat{A}$ by itself is the event that $\xi_{n,1} = j$ and the first j in X_{n+1} is reached by a jump from i. See (76) and Figure 7.

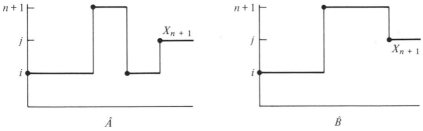

Figure 7.

(78) Lemma. *Let $n \geq 2$. Fix $i \neq j$ in I_n. Given $\xi_{n,1} = j$, the conditional* $\mathbf{P_i}$-*probability that the first j in X_{n+1} was reached by a jump from i is* $Q_{n+1}(i,j)/Q_n(i,j)$.

PROOF OF (77).   Relation (74) implies:

$$(79) \qquad \mathbf{P_i}\{M = m\} = \mathbf{P_i}\{A_m \cup B_m\} = [1 - \pi_{n+1}(i)]^m \, \pi_{n+1}(i);$$

$$(80) \quad \mathbf{P_i}\{\tau_{n+1,2\nu} > s_\nu \text{ for } \nu = 0,\ldots,m \text{ and } M = m\} = \mathbf{P_i}\{M = m\}e^{-s},$$

where $s_0, s_1, \ldots, s_m$ are nonnegative numbers, and

$$s = q_{n+1}(i) \cdot \Sigma_{\nu=0}^m \, s_\nu.$$

Now use (71), with $q = q_{n+1}(i)$ and $F = \{0, 1\}$ and $p(1) = \pi_{n+1}(i)$. Construct independent objects $Z_1, Z_2, \ldots, \mathscr{P}$, where: each Z_m is 1 with probability $p(1)$ and 0 with probability $p(0) = 1 - p(1)$; and $\mathscr{P} = (T_0, T_1, \ldots)$ is a Poisson process with parameter q. Let N be the least m if any with $Z_m = 1$, and $N = \infty$ if none. By (75, 79, 80), the distribution of $(T_0, \ldots, T_{N-1}, T_N)$ coincides with the $\mathbf{P_i}$-distribution of $(\theta_0, \ldots, \theta_{M-1}, \tau_{n,0})$. Remember that $\mathscr{P}_f$ is the set of T_n with $Z_n = f$. Of course, T_N is the first time in $\mathscr{P}_1$, and $(T_0, T_1, \ldots, T_{N-1})$ is the restriction of $\mathscr{P}_0$ to $(0, T_N)$. But (71) makes T_N exponentially distributed with parameter $\pi_{n+1}(i)q_{n+1}(i)$, and $\mathscr{P}_0$ Poisson with parameter $[1 - \pi_{n+1}(i)]q_{n+1}(i)$, independent of T_N.   ★

PROOF OF (73a).   This is part of (77).   ★

PROOF OF (73b, c) AND (78).   Suppose $n \geq 2$. Fix $j \neq i$ in I_n. Temporarily, let $g = \Gamma_{n+1}(i,j)$ and $d = \Gamma_{n+1}(i, n + 1)\Gamma_{n+1}(n + 1, j)$ and $p = \pi_{n+1}(i)$. Then (74) makes $\mathbf{P_i}(\hat{A}_m) = (1 - p)^m g$, so $\mathbf{P_i}(\hat{A}) = g/p$. Similarly, $\mathbf{P_i}(\hat{B}) = d/p$. Thus,

$$\Gamma_n(i,j) = \mathbf{P_i}(\hat{A} \cup \hat{B}) = (g + d)/p,$$

proving (73b). Then (73c) follows on multiplying by $q_{n+1}(i)$ and using (73a). The conditional probability in (78) is $\mathbf{P_i}(\hat{A}|\hat{A} \cup \hat{B}) = g/(g + d)$, which simplifies on using (73).   ★

The rest of this section completes the specification of the conditional $\mathbf{P_i}$-distribution of X_{N+1} given X_N; notation has been changed here a little, to provide an extra running index n. Fix N and i in I_N. Confine ω to $\{X(0) = i\}$. As (35) implies, X_N is obtained from X_{N+1} by erasing the $(N + 1)$-intervals. In particular, X_N is a function of X_{N+1}. Conversely, there are random cut times $T_0, T_1, \ldots$ with $0 < T_0 < T_1 < \ldots < T_n \to \infty$, and random lengths $L_0, L_1, \ldots$ which are positive and have infinite sum, such that X_{N+1} is obtained by cutting X_N at times $T_0, T_1, \ldots$ and inserting $(N + 1)$-intervals of lengths $L_0, L_1, \ldots$. That is,

$X_{N+1}(t) = X_N(t)$

    for $0 \leqq t < T_0$

$= N + 1$

    for $T_0 \leqq t < T_0 + L_0$

$\vdots$

$= X_N(t - L_0 - \cdots - L_n)$

    for $T_n + L_0 + \cdots + L_n \leqq t < T_{n+1} + L_0 + \cdots + L_n$

$= N + 1$

    for $T_{n+1} + L_0 + \cdots + L_n \leqq t < T_{n+1} + L_0 + \cdots + L_{n+1}$,

as in Figure 8.

Of course, L_n is the length of the nth $(N + 1)$-interval in X_{N+1}, for $n = 0, 1, \ldots$; and T_n is the time on the X_N-scale corresponding to any time on the X_{N+1}-scale in the nth $(N + 1)$-interval. The problem is to obtain the conditional $\mathbf{P_i}$-distribution of the T's and L's, given X_N. The L's are easy.

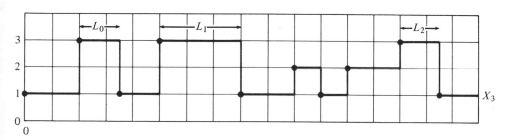

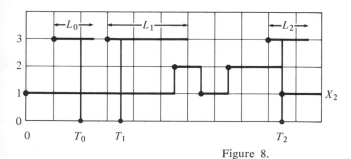

Figure 8.

(81) Theorem. *Fix $i \in I_N$. Relative to $\mathbf{P}_i$:*

(a) *$(L_0, L_1, \ldots)$ is independent of $(T_0, T_1 \ldots, X_N)$;*

(b) *$L_0, L_1, \ldots$ are independent and exponential with common parameter $q_{N+1}(N + 1)$.*

PROOF. Use (74). The σ-field spanned by $(T_0, T_1, \ldots, X_N)$ coincides with the σ-field Σ spanned by $\xi_{N+1,n}$ for all n, and by $\tau_{N+1,n}$ for those n with $\xi_{N+1,n} \neq N + 1$. And $(L_0, L_1, \ldots)$ is the sequence of $\tau_{N+1,n}$ for those n with $\xi_{N+1,n} = N + 1$. To crisp this up, let $U_0, U_1, \ldots$ be the sequence of $\tau_{N+1,n}$ for those n with $\xi_{N+1,n} \neq N + 1$. Let A be the event

$$L_0 > t_0, L_1 > t_1, \ldots, L_a > t_a$$

for nonnegative $t_0, t_1, \ldots, t_a$. Let B be the event

$$U_0 > u_0, U_1 > u_1, \ldots, U_b > u_b$$

$$\xi_{N+1,0} = i_0, \xi_{N+1,1} = i_1, \ldots, \xi_{N+1,c} = i_c,$$

where: $u_0, u_1, \ldots, u_b$ are nonnegative; $i_0 = i$ and $i_1, \ldots, i_c$ are in I_{N+1}: the sequence $i_0, i_1, \ldots, i_c$ has at least $a + 1$ terms equal to $N + 1$ and at least $b + 1$ terms different from $N + 1$. Let

$$s = q_{N+1}(N + 1)[t_0 + t_1 + \cdots + t_a].$$

By (74):

(82) $$\mathbf{P}_i\{A \cap B\} = e^{-s}\,\mathbf{P}_i\{B\}.$$

Put $b = -1$, so B does not constrain the U's. Sum over all possible sequences $i_0, i_1, \ldots, i_c$ which have exactly $a + 1$ terms equal to $N + 1$, and $i_0 = i$ and $i_c = N + 1$; even c is allowed to vary. Because ξ_{N+1} visits $N + 1$ at least $a + 1$ times, the union of the B's is $\{X_N(0) = i\}$, and different B's are disjoint. So

(83) $$\mathbf{P}_i\{A\} = e^{-s}.$$

Agree the null set is a B: this doesn't hurt (82). As $b, c, u_0, \ldots, u_b, i_0, \ldots, i_c$ vary, the class of B's is closed under intersection and generates Σ: for ω is confined to Ω_∞. So (82) holds for all $B \in \Sigma$, by $(MC, 10.16)$. A similar effort on A completes the argument.   ★

Next, suppose $N \geq 2$. Look at Figure 8 again. There are two kinds of cuts: the first kind appears at a jump of X_N, and the second interior to an interval of constancy for X_N. Let $\sigma_1 = \tau_{N,0}$, $\sigma_2 = \tau_{N,0} + \tau_{N,1}, \ldots$ and so on, the times of the discontinuities of X_N, numbered from 1 up, as in Figure 9.

Let C_n be the event that some $T_m = \sigma_n$; that is, there is a cut at the nth discontinuity of X_N. Let V_n be a point process on $(0, \infty)$; namely, V_0 is the

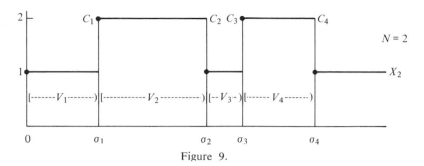

Figure 9.

set of times T_m such that $T_m < \sigma_1$; and V_n is the set of times $T_m - \sigma_n$ such that $\sigma_n < T_m < \sigma_{n+1}$. Thus, V_n is the set of cuts occurring interior to the nth interval of constancy (σ_n, σ_{n+1}) for X_N, referred to σ_n as the new origin of time. In particular, the number of times in V_n is random, finite and may well be 0.

(84) Theorem. *Suppose $M = 0, 1, \ldots$. Given $\xi_{N,0}, \xi_{N,1}, \ldots, \xi_{N,M}$:*
  (a) $(\tau_{N,0}, V_0), C_1, (\tau_{N,1}, V_1), C_2, \ldots (\tau_{N,M}, V_M), C_{M+1}$ *are conditionally* $\mathbf{P_j}$-*independent*;
  (b) *for $1 \leq m \leq M$, the conditional $\mathbf{P_j}$-distribution of $(\tau_{N,m}, V_m)$ on* $\{\xi_{N,m} = j\}$ *coincides with the unconditional $\mathbf{P_j}$-distribution of $(\tau_{N,0}, V_0)$;*
  (c) *for $1 \leq m \leq M$, the conditional $\mathbf{P_j}$-probability of C_m on the set* $\{\xi_{N,m-1} = j, \xi_{N,m} = k\}$ *coincides with the $\mathbf{P_j}$-probability of C_1 given $\{\xi_{N,1} = k\}$.*

  PROOF. Use the finite I case of (109) below.   ★

The conditional $\mathbf{P_j}$-probability of C_1 given $\{\xi_{N,1} = k\}$ was obtained in (78), and is

$$1 - [Q_{N+1}(j, k)/Q_N(j, k)].$$

The $\mathbf{P_j}$-distribution of $(\tau_{N,0}, V_0)$ was obtained in (77), and coincides with the distribution of τ and the restriction of $\mathscr{P}$ to $(0, \tau)$, where τ and $\mathscr{P}$ are independent, τ is exponential with parameter

$$q_N(j) = \pi_{N+1}(j)q_{N+1}(j),$$

and $\mathscr{P}$ is Poisson with parameter

$$q_{N+1}(j) - q_N(j) = [1 - \pi_{N+1}(j)]q_{N+1}(j).$$

This theorem can be put in integrated form, just as (66) was.

(85) Summary. **(a)** X_{N+1} is obtained from X_N by cutting X_N and inserting $(N + 1)$-intervals. Given X_N and the locations of the cuts, the lengths of the inserted intervals are independent and exponential, with common parameter $q_{N+1}(N + 1)$. There are two kinds of cuts: the first kind appears at a jump of X_N, and the second kind appears interior to an interval of constancy for X_N. Cuts of the first kind appear independently from jump to jump, and independently of cuts of the second kind. Locations of cuts of the second kind are independent from interval to interval. At a jump from j to k, the probability of a cut not appearing is $Q_{N+1}(j, k)/Q_N(j, k)$. Within a particular j-interval, the location of cuts has a Poisson distribution, with parameter $q_{N+1}(j) - q_N(j)$.

(b) The conditional distribution of X_{N+m+1} given $X_{N+m}, \ldots, X_N$ coincides with the conditional distribution of X_{N+m+1} given X_{N+m}, because X_{N+m} determines $X_{N+m-1}, \ldots, X_N$. Thus, (a) can be iterated to give the conditional distribution of X_{N+m} given X_N.

On a first reading of the book, skim to (87).

(86) Cutting and Pasting. Suppose a process $\hat{X}_{N+1}$ is obtained from X_N by the procedure described in (85), except the probability of a cut appearing at a jump from j to k is a given number $p(j, k)$, and the location of cuts in a j-interval has a given parameter $p(j)$. A little experimentation shows that $\hat{X}_{N+1}$ need not be Markov. For instance, let $N = 3$ and $p(1, 2) = p(2, 3) = 1$, while all other p's vanish. If $\hat{X}_{N+1}$ were Markov, then jumps $1 \to 4$ and $4 \to 3$ would be possible; so the sequence $1 \to 4 \to 3$ would be possible. But it isn't. Suppose, however, that there is a standard stochastic semigroup $\hat{P}_{N+1}$ on I_{N+1}, whose generator $\hat{Q}_{N+1} = \hat{P}'_{N+1}(0)$ gives Q_N when restricted to I_N. This condition is easy to check from (73). Suppose further that

$$p(j, k) = 1 - [\hat{Q}_{N+1}(j, k)/Q_N(j, k)]$$

and

$$p(j) = |\hat{Q}_{N+1}(j, j)| - q_N(j).$$

Then indeed $\hat{X}_{N+1}$ is Markov, with transitions $\hat{P}_{N+1}$. This follows from (85a) and a general principle. Suppose (X, Y) and $(\hat{X}, \hat{Y})$ are pairs of random objects. Suppose $\hat{X}$ is distributed like X. Suppose $D(x)$ is a regular conditional distribution both for $\hat{Y}$ given $\hat{X} = x$ and for Y given $X = x$. Then $(\hat{X}, \hat{Y})$ is distributed like (X, Y), and $\hat{Y}$ is distributed like Y.

Abbreviate

$$s(j) = \Sigma_k \{p(j, k)Q_N(j, k) : k \in I_N \text{ and } k \neq j\}.$$

The condition on p that makes $\hat{X}_{N+1}$ Markov can be restated as follows: there is a probability θ on I_N such that for all $j \neq k$ in I_N:

$$p(j, k)Q_N(j, k) = [s(j) + p(j)]\theta(k).$$

If p comes from $\hat{Q}_{n+1}$ then $\theta(k) = \hat{Q}_{N+1}(N + 1, k)/|\hat{Q}_{N+1}(N + 1, N + 1)|$; and the condition holds by (73) and algebra. If p satisfies the condition, you can define $\hat{Q}_{N+1}$ as follows: For $j \neq k$ in I_N,

$$\hat{Q}_{N+1}(j, k) = [1 - p(j, k)]Q_N(j, k);$$

for j in I_N,

$$\hat{Q}_{N+1}(j, N + 1) = s(j) + p(j)$$

$$\hat{Q}_{N+1}(j, j) = -q_N(j) - p(j)$$

$$\hat{Q}_{N+1}(N + 1, j) = \hat{q}_{N+1}(N + 1)\theta(j);$$

and

$$\hat{Q}_{N+1}(N + 1, N + 1) = -\hat{q}_{N+1}(N + 1) \quad \text{is free.}$$

You have to check that $\hat{Q}_{N+1}$ is a generator, using $(MC, 5.29)$; that $(\hat{Q}_{N+1})_{I_N} = Q_N$, using (73); and that the instructions are right, using (85a). You might find it helpful to observe

$$p(j) = [s(j) + p(j)]\theta(j):$$

sum out $k \in I_N \setminus \{j\}$ in the condition, and rearrange. To keep I_{N+1} a recurrent class, you need

$$s(j) + p(j) > 0 \quad \text{for some } j.$$

Suppose Q_N is given. Here† is a canonical way to generate θ and p. Choose any probability θ on I_N. Fix $j \in I_N$. Suppose $\theta(j) > 0$. Choose any finite $p(j)$ with

$$0 \leq p(j) \leq \inf_h \theta(j)Q_N(j, h)/\theta(h):$$

where $h \in I_N \setminus \{j\}$ and

$$\theta(j)Q_N(j, h)/\theta(h) = \infty$$

when $\theta(h) = 0$, even if $Q_N(j, h) = 0$. Then

$$p(j, k) = \frac{\theta(k)p(j)}{\theta(j)Q_N(j, k)} \quad \text{when } Q_N(j, k) > 0.$$

Suppose $\theta(j) = 0$. Then $p(j) = 0$. If $\theta(h) > 0$ and $Q_N(j, h) = 0$ for any h, then

$$p(j, k) = 0 \quad \text{when } Q_N(j, k) > 0.$$

If $\theta(h) > 0$ implies $Q_N(j, h) > 0$, choose any $s(j)$ with

$$0 \leq s(j) \leq q_N(j).$$

† I will say what the facts are, but won't argue them. On a first reading, don't stop to think.

Then

$$p(j, k) = s(j)\theta(k)/Q_N(j, k) \quad \text{when } Q_N(j, k) > 0.$$

You might check

$$s(j) = \Sigma_k \{p(j, k)Q_N(j, k) : k \in I_N \text{ and } k \neq j\}.$$

If $Q_N(j, k) = 0$, then $p(j, k)$ is relatively meaningless: choose it in $[0, 1]$. ★

The time spent in i until the tth instant in j

The next result will be useful in Chapter 3. Let $i \neq j$ be in I_N. Let $\pi_N(i, j)$ be the probability that a discrete time Markov chain which starts from i and moves according to Γ_N visits j before returning to i: so $\pi_N(i, j) > 0$. Let $W_{i,j}(t)$ be the time X_N spends in j, until X_N has spent time t in i. So $W_{i,j}$ doesn't depend on N. I will formalize this, using the notation of Section 4. Composition is in the time domain, and is done separately for each ω.

$$W_{i,j} = \mu_{\{j\},I_N} \circ \gamma_{\{i\},I_N}$$

$$= \mu_{\{j\}} \circ \gamma_{I_N} \circ \mu_{I_N} \circ \gamma_{\{i\}} \quad \text{(32a, c)}$$

$$= \mu_{\{j\}} \circ \gamma_{\{i\}} \quad \text{(25d, f).}$$

(87) Theorem. $W_{i,j}(t) = \Sigma_{n=1}^{Z(t)} Y_n$, where relative to $\mathbf{P_i}$:
   (a) $(Z, Y_1, Y_2, \ldots)$ are independent;
   (b) Z is a Poisson process with parameter $\pi_N(i, j)q_N(i)$;
   (c) Y_n have common exponential distribution, with parameter $\pi_N(j, i)q_N(j)$.

So

   (d) $$\int W_{i,j}(t) \, d\mathbf{P_i} = \frac{\pi_N(i, j)q_N(i)}{\pi_N(j, i)q_N(j)} t.$$

PROOF. If $N = 2$, then $\pi_N = 1$; and (87) follows from $(MC, 5.48)$. Suppose $N > 2$. By renumbering, suppose $i = 1$ and $j = 2$. As for (73a), or from (87e),

$$q_2(1) = \pi_N(1, 2)q_N(1) \quad \text{and} \quad q_2(2) = \pi_N(2, 1)q_N(2).$$

Now use the case $N = 2$ on X_2; remember that $W_{1,2}$ doesn't depend on N. ★

More generally, let $J \subset K \subset H$ be finite subsets of I. Let $\pi(i, J, K)$ be the probability that a discrete-time chain starting from i and moving according to Γ_K hits $J\setminus\{i\}$ before returning to i. Then

(87e) $$\pi(i, J, H) = \pi(i, J, K) \cdot \pi(i, K, H)$$

(87f) $$q_J(i) = \pi(i, J, K) \cdot q_K(i).$$

To prove (87e), let $r = 1 - \pi(i, K, H)$, the probability that a Γ_H-chain starting from i returns to i before hitting $K \setminus \{i\}$. By strong Markov, $r^n\pi(i, J, H)$ is the probability that a Γ_H-chain starting from i experiences n returns to i without hitting $K \setminus \{i\}$, and then hits $J \setminus \{i\}$ before the next return to i. So

$$\Sigma_{n=0}^{\infty} r^n\pi(i, J, H)$$

is the probability that this same chain, restricted to K, hits $J \setminus \{i\}$ before returning to i, which is $\pi(i, J, K)$. This proves (87e). Then (87f) follows by iterating (73a) and using (87e), because

$$\pi_{n+1}(i) = \pi(i, I_n, I_{n+1}).$$

Otherwise, (87f) can be argued like (73a).

10. THE CONVERGENCE OF Q_J TO Q

Continue in the setting of Section 5. Remember that X is Markov with transitions P, and $Q = P'(0)$. The restriction X_J of X to J is Markov with transitions P_J, and $Q_J = P'_J(0)$. The main results of this section are:

(88) Theorem. $Q_J(i, i) \downarrow Q(i, i)$ as $J \uparrow I$ for each i in I.

(89) Theorem.   For each $i \neq j$ in I and $\varepsilon > 0$, there is a $\delta > 0$ such that for all $t \leq \delta$ and finite $J \supset \{i, j\}$,

   (a) $P_J(t, i, j) \geq (1 - \varepsilon)P(t, i, j)$.

Moreover, there is a finite J with $\{i, j\} \subset J$ and a positive δ such that for all $t \leq \delta$ and finite $K \supset J$,

   (b) $P_K(t, i, j) \leq (1 + \varepsilon)P(t, i, j) + \varepsilon t$.

If $Q(i, j) > 0$, the second term on the right in (b) is redundant. If $Q(i, j) = 0$, the first term on the right in (b) is redundant, but the second term is vital: for $Q_K(i, j)$ is likely to be positive. The analog of (89) for $i = j$ fails:

$$(1 - \varepsilon)[1 - P(t, i, i)] \leq 1 - P_J(t, i, i) \quad \text{for } t \leq \tfrac{1}{17}$$

is impossible when i is instantaneous.

(90) Corollary. $Q_J(i, j) \downarrow Q(i, j)$ as $J \uparrow I$ for all i, j in I.

(91) Corollary. $P'(0, i, j)$ exists and is finite for all $i \neq j$ in I.

This was proved somewhat more efficiently as $(MC, 5.21)$.

PROOF OF (88).   Suppose $i \in J$ and confine ω to $\{X(0) = i\}$. Let τ be the first holding time of X, and let τ_J be the first holding time of X_J. I say $\tau_J \downarrow \tau$ as $J \uparrow I$. Indeed, $\gamma_J(\tau_J)$ is the least t with $X(t) \in J \setminus \{i\}$, by (28). And τ is the inf of t with $X(t) \in I \setminus \{i\}$. So $\gamma_J(\tau_J) \downarrow \tau$ as $J \uparrow I$. Check

$$\tau = \mu_J(\tau) \leq \mu_J[\gamma_J(\tau_J)] = \tau_J.$$

This proves $\tau \leq \tau_J$. Now (35) makes τ_J nonincreasing with J. Use (25b):

$$\tau \leq \tau_J \leq \gamma_J(\tau_J).$$

This completes the proof that $\tau_J \downarrow \tau$ as $J \uparrow I$. Now use $(MC, 9.18)$.   ★

Isaac Meilijson put the final touch on this proof of (88).

(92) Lemma. *Fix $j \in I$. Let f be the least t with $X(t) = j$. Then*

$$P(t,i,j) = P_i\{f \leq t \text{ and } X(t) = j\} = \int_{\{f \leq t\}} P(t - f, j, j) \, dP_i.$$

NOTE. $\{X(t) = j\} \subset \{f \leq t\}$.

PROOF.   Confirm that f is Markov, and $X(f) = j$. Use the strong Markov property. Let W be the set of $\omega \in \Omega_\infty$ such that $X[f(\omega) + \cdot, \omega]$ is quasiregular. Then $P_i(W) = 1$ by $(MC, 9.41a)$. Let $S(\omega)$ be $X[f(\omega) + \cdot, \omega]$ retracted to the binary rationals R. Let

$$F(\omega, \omega') = 1 \quad \text{when } f(\omega) \leq t \text{ and } X[t - f(\omega), \omega'] = j$$
$$= 0 \quad \text{elsewhere.}$$

Let $\mathscr{F}$ be the product σ-field in Ω relativized to Ω_∞. But f is $\mathscr{F}(f +)$-measurable, so F is $\mathscr{F}(f +) \times \mathscr{F}$-measurable on $\Omega_\infty \times \Omega_\infty$. Make sure that $F(\cdot, S)$ coincides with the indicator of $\{f \leq t \text{ and } X(t) = j\}$, at least on W. Now use $(MC, 9.41e)$:

$$P_i\{f \leq t \text{ and } X(t) = j\} = P_i\{W \text{ and } f \leq t \text{ and } X(t) = j\}$$

$$= \int_W F(\cdot, S) \, dP_i$$

$$= \int_W F^* \, dP_i.$$

If $f(\omega) \leq t$, then

$$F^*(\omega) = \int F(\omega, \omega') \, P_j(d\omega')$$

$$= P_j\{\omega' : X[t - f(\omega), \omega'] = j\}$$
$$= P(t - f(\omega), j, j).$$

If $f(\omega) > t$, then $F^*(\omega) = 0$.   ★

PROOF OF (89).   Suppose $0 < \varepsilon < 1$. Fix $i \neq j$ in I. Let J and K be finite subsets of I, with $\{i, j\} \subset J \subset K$. Confine ω to $\{X(0) = i\}$. Introduce f_J, the least t with $X_J(t) = j$. By (47), there is a positive δ such that

(93) $P_J(s, j, j) \geq 1 - \varepsilon$ for $0 \leq s \leq \delta$ and all J.

Use (92) on the distribution of X_J, which is legitimate by (36–37):

$$P_i\{f_J \leq t \text{ and } X_J(t) = j\} = \int_{\{f_J \leq t\}} P_J(t - f_J, j, j) \, dP_i.$$

Use (93):

(94) $P_i\{f_J \leq t \text{ and } X_J(t) \neq j\} \leq \varepsilon P_i\{f_J \leq t\}$ for $0 \leq t \leq \delta$.

Remember $\{i, j\} \subset J \subset K \subset I$. I say

$$\{X_K(t) = j\} \subset \{X_J[\mu_{J,K}(t)] = j\}.$$

Indeed, (32a) makes

$$X_J[\mu_{J,K}(t)] = X[\gamma_J(\mu_J(\gamma_K(t)))].$$

But $X_K(t) = X[\gamma_K(t)] = j$ makes $\gamma_K(t) \in S_J$, so (25f) makes

$$\gamma_J(\mu_J(\gamma_K(t))) = \gamma_K(t);$$

forcing

$$X_J[\mu_{J,K}(t)] = X[\gamma_K(t)] = X_K(t) = j.$$

Next, $\mu_{J,K}(t) \leq t$. So

$$\{X_K(t) = j\} \subset \{f_J \leq t\} \subset \{X_J(t) = j\} \cup \{f_J \leq t \text{ and } X_J(t) \neq j\}.$$

This and (94) give

$$P_K(t, i, j) \leq P_i\{f_J \leq t\} \leq P_J(t, i, j) + \varepsilon P_i\{f_J \leq t\}   \text{ for } 0 \leq t \leq \delta,$$

so

(95) $P_J(t, i, j) \geq (1 - \varepsilon)P_i\{f_J \leq t\} \geq (1 - \varepsilon)P_K(t, i, j)$ for $0 \leq t \leq \delta$.

Let K increase to I in (95) and use (53a) to get (89a).
In particular, (89a) and (95) imply

$$Q(i, j) \leq Q_K(i, j) \leq Q_J(i, j).$$

This also follows from (73c). Let

$$Q_I(i, j) = \lim_{J \uparrow I} Q_J(i, j).$$

Plainly,

$$Q_I(i,j) \geq Q(i,j).$$

Turn now to (89b). The case $Q_I(i,j) = 0$ is easy; for then $Q(i,j) = 0$. Moreover, $Q_J(i,j) \leq \frac{1}{2}\varepsilon(1 - \varepsilon)$ for some large J. Then there is a $\delta_J > 0$, such that

(96) $P_J(t,i,j) \leq Q_J(i,j)t + \frac{1}{2}\varepsilon(1 - \varepsilon)t$ for $0 \leq t \leq \delta_J$.

From (95), with $0 \leq t \leq \min\{\delta, \delta_J\}$ and $K \supset J$,

$$P_K(t,i,j) \leq \frac{1}{1 - \varepsilon} P_J(t,i,j)$$

$$\leq \frac{1}{1 - \varepsilon}[Q_J(i,j)t + \frac{1}{2}\varepsilon(1 - \varepsilon)t]$$

$$\leq \varepsilon t.$$

For the case $Q_I(i,j) > 0$, recall that X_J visits $\xi_{J,0}, \xi_{J,1}, \ldots$ with holding times $\tau_{J,0}, \tau_{J,1}, \ldots$. As in Figure 10, let

(97) $A_J = \{\xi_{J,0} = i \text{ and } \xi_{J,1} = j\}$

$$A_J(t) = \{A_J \text{ and } \tau_{J,0} \leq t < \tau_{J,0} + \tau_{J,1}\}.$$

Let u and v be nonnegative numbers. As (37) and $(MC, 5.48)$ imply:

(98) $P_i\{A_J \text{ and } \tau_{J,0} > u \text{ and } \tau_{J,1} > v\} = \Gamma_J(i,j)\, e^{-s}$,   where

$$s = q_J(i)u + q_J(j)v.$$

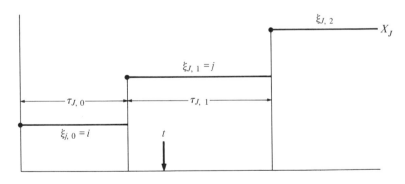

$$A_J(t)$$

Figure 10.

For corollary (98*), let U_i and U_j be independent, exponential random variables, with parameters $q_J(i)$ and $q_J(j)$, respectively. Abbreviate $b = q_J(j) - q_J(i)$ and $c = q_J(j)$. For $0 \leq s \leq t$, Fubini and (98) show:

(98*) $P_i\{\tau_{J,0} \leq s \text{ and } A_J(t)\} = \Gamma_J(i, j) \cdot \text{Prob}\{U_i \leq s \leq t < U_i + U_j\}$

$$= Q_J(i, j) \cdot e^{-ct} \cdot \int_0^s e^{bu}\, du$$

$$= Q_J(i, j) \cdot [s + o(t)] \quad \text{as } t \to 0.$$

Remember that f is the least t with $X(t) = j$. The main thing to see is

(99) If $Q_I(i, i) > 0$ and $0 < a < 1$, there is a positive d, and a finite subset J of I which includes $\{i, j\}$, for which

$$P_i\{f \leq t \text{ and } A_J(t)\} \geq (1 - a)P_i\{A_J(t)\} \quad \text{when } 0 \leq t \leq d.$$

I will prove this later. Suppose it for the moment, and suppose $Q_I(i, j) > 0$. I will argue (89b). For some positive $d_1 \leq d$,

$$P_i\{A_J(t)\} \geq (1 - a)Q_J(i, j)t \quad \text{when } 0 \leq t \leq d_1;$$

this follows from (98*) with $s = t$, taking into account the relation

(100) $Q_J(i, j) \geq Q_I(i, j) > 0.$

Thus

$$P_i\{f \leq t\} \geq (1 - a)^2 Q_J(i, j)t \qquad \text{when } 0 \leq t \leq d_1.$$

For some positive $d_2 \leq d_1$,

$$(1 - a)P_J(t, i, j) \leq Q_J(i, j)t \qquad \text{when } 0 \leq t \leq d_2,$$

because $P'_J(0, i, j) = Q_J(i, j) > 0$ by (100). Thus,

$$P_i\{f \leq t\} \geq (1 - a)^3 P_J(t, i, j) \quad \text{when } 0 \leq t \leq d_2.$$

Find a positive $d_3 \leq d_2$ so that

$$P(t, j, j) \geq 1 - a \qquad \text{when } 0 \leq t \leq d_3.$$

By (92),

$$P(t, i, j) \geq (1 - a)P_i\{f \leq t\} \qquad \text{when } 0 \leq t \leq d_3.$$

Thus,

$$P(t, i, j) \geq (1 - a)^4 P_J(t, i, j) \qquad \text{when } 0 \leq t \leq d_3.$$

According to (95), there is a positive $d_4 \leq d_3$ such that

$$P_J(t, i, j) \geq (1 - a)P_K(t, i, j) \quad \text{when } 0 \leq t \leq d_4 \text{ and } K \supset J.$$

Thus

$$P(t, i, j) \geq (1 - a)^5 P_K(t, i, j) \quad \text{when } 0 \leq t \leq d_4 \quad \text{and } K \supset J.$$

Here d_4 depends on i, j, and J, but does not depend on K. This completes the derivation of (89b) from (99), in the case $Q_J(i, j) > 0$.

I still have to prove (99). Here are some preliminaries. Do not assume $Q_J(i, j) > 0$ in proving (101–102). Do assume $Q_J(i, j) > 0$. To state (101), let E_K be the event that the first j in X_K is reached by a jump from i, as in Figure 11. I claim

(101) $$P_i\{E_K \text{ and } H\} = \frac{Q_K(i, j)}{Q_J(i, j)} \cdot P_i\{H\},$$

for all subsets H of E_J which are measurable on X_J. To begin with, suppose $K = J \cup \{k\}$ with $k \notin J$. Let $0 \leq m < M$ be integers. Let $i_0, i_1, \ldots, i_n$ be a J-sequence, such that $i_0 = i_m = i$ and $i_{m+1} = j$ and $i_n \neq j$ for $0 < n < M$

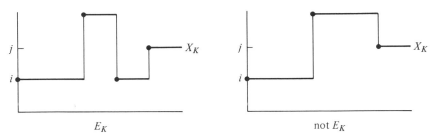

$$E_K \qquad\qquad\qquad \text{not } E_K$$

Figure 11.

Let $t_0, \ldots, t_M$ be nonnegative numbers. Let

$$H = \{\tau_{J,n} > t_n \text{ and } \xi_{J,n} = i_n \text{ for } n = 0, \ldots, M\}.$$

Let C_n be the event that the nth discontinuity in X_J is cut in passing to X_K. So

$$\{H \text{ and } E_K\} = \{H \text{ but not } C_{m+1}\}$$

Use (84) and (78):

$$P_i\{H \text{ and } E_K\} = P_i\{H \text{ but not } C_{m+1}\}$$
$$= P_i\{\text{not } C_1 | \xi_{J,1} = j\} \cdot P_i\{H\}$$
$$= \frac{Q_K(i, j)}{Q_J(i, j)} \cdot P_i\{H\}.$$

Now $H \subset E_J$. The class of special H is closed under intersection, modulo the null set, and generates the same σ-field on E_J as X_J. Use $(MC, 10.16)$ to complete the proof of (101), when $K = J \cup \{k\}$ with $k \notin J$. To prove general (101), adopt the notation of Section 9. Number the states so $I = \{1, 2, \ldots\}$ and $J = I_N = \{1, 2, \ldots, N\}$ and $K = I_M$ with $M > N$. Abbreviate X_n for X_{I_n} and Q_n for Q_{I_n} and E_n for E_{I_n}. Remember that P_i is the probability which makes X Markov with stationary transitions P and starting state i. Let $N \leq m < M$. I say

$$P_i\{E_{m+1} \cap H\} = P_i\{E_{m+1} \cap E_m \cap H\}$$

$$= \frac{Q_{m+1}(i, j)}{Q_m(i, j)} \cdot P_i\{E_m \cap H\}:$$

the first equality holds because $E_{m+1} \subset E_m$; the second holds by the special case of (101), because $E_m \cap H$ is measurable on X_m. Now induction settles general (101). Let F_J be the intersection of E_K, as K varies through the finite subsets of I which include J. Let $K \uparrow I$ in (101), so $E_K \downarrow F_J$, and conclude

(102) $$P_i\{F_J \text{ and } H\} = \frac{Q_I(i, j)}{Q_J(i, j)} \cdot P_i\{H\},$$

for all subsets H of E_J which are measurable on X_J.

For the rest, suppose $Q_I(i, j) > 0$. Choose J so large that

(103) $$\frac{Q_I(i, j)}{Q_J(i, j)} > 1 - \tfrac{1}{3}a.$$

Remember definition (97) of A_J and $A_J(t)$; these are X_J-measurable subsets of E_J. Put $A_J(t)$ for H in (102) and use (103):

(104) $P_i\{F_J \text{ and } A_J(t)\} > (1 - \tfrac{1}{3}a)P_i\{A_J(t)\}$ for all $t \geq 0$.

Recall that $\gamma_J(t-)$ is the least and $\gamma_J(t)$ is the greatest X-time corresponding to X_J-time t. Now $\gamma_J(\tau_{J,0}) = \alpha_{J,0} + \beta_{J,1}$, where $\alpha_{J,0} = \gamma_J(\tau_{J,0}-)$ is the amount of time X spends interior to the initial interval of X_J-constancy, and $\beta_{J,1}$ is the time X spends at the first jump in X_J. I say

(105) $$\beta_{J,1} = 0 \quad \text{on } F_J \cap A_J.$$

Indeed,

$$\beta_{J,1} = \gamma_J(\tau_{J,0}) - \gamma_J(\tau_{J,0}-)$$

$$= \lim_{K \uparrow I} [\gamma_{J,K}(\tau_{J,0}) - \gamma_{J,K}(\tau_{J,0}-)] \quad \text{by (34).}$$

But $\gamma_{J,K}(\tau_{J,0}) = \gamma_{J,K}(\tau_{J,0}-)$ on $E_K \cap A_J$. This proves (105). From (104) and (105),

(106) $P_i\{\beta_{J,1} = 0 \text{ and } A_J(t)\} > (1 - \tfrac{1}{3}a)P_i\{A_J(t)\}$ for all $t \geq 0$.

Fix a with $0 < a < 1$. Specialize $s = (1 - \frac{1}{4}a)t$ and $s = t$ in (98*); remember $\tau_{J,0} \leqq t$ on $A_J(t)$. There is a positive d_0 for which

(107) $P_i\{\tau_{J,0} \leqq (1 - \frac{1}{4}a)t$ and $A_J(t)\} > (1 - \frac{1}{3}a)P_i\{A_J(t)\}$ when $0 \leqq t \leqq d_0$.

Using (56) and (66), with the emphasis on (56e), I will produce a positive $d \leqq d_0$ such that

(108) $P_i\{\alpha_{J,0} \leqq (1 + \frac{1}{4}a)\tau_{J,0}$ and $A_J(t)\} \geqq (1 - \frac{1}{3}a)P_i\{A_J(t)\}$ when $0 \leqq t \leqq d$.

Indeed, $(\tau_{J,0}, \alpha_{J,0})$ and $\tau_{J,1}$ are conditionally P_i-independent given A_J, by (66). Use (56). On a convenient probability triple, there exist independent random objects G, T, S, where: the process G satisfies (56e); while S and T are random variables; and the joint distribution of $(T, G(T-), S)$ coincides with the joint conditional P_i-distribution of $(\tau_{J,0}, \alpha_{J,0}, \tau_{J,1})$ given A_J. The left side of (108) is then $\Gamma_J(i, j)p$, where

$$p = \text{Prob}\,\{G(T-) \leqq (1 + \tfrac{1}{4}a)T \ \text{and}\ T \leqq t < T + S\}.$$

And $P_i\{A_J(t)\} = \Gamma_J(i, j)q$, where

$$q = \text{Prob}\,\{T \leqq t < T + S\}.$$

Let m be the joint distribution of T and S, a probability on the positive quadrant. By Fubini,

$$p = \int_{\substack{0 < u, v < \infty \\ u \leqq t < u + v}} \text{Prob}\,\{G(u-) \leqq (1 + \tfrac{1}{4}a)u\}\, m(du, dv).$$

Property (56e) provides a positive $d \leqq d_0$ such that the kernel of this integral is at least $(1 - \frac{1}{3}a)$ for all $u \leqq d$, even if $u-$ is replaced by u. Therefore, $p \geqq (1 - \frac{1}{3}a)q$, proving (108).

Recall $\gamma_J(\tau_{J,0}) = \alpha_{J,0} + \beta_{J,1}$. If $\beta_{J,1} = 0$ and $\tau_{J,0} \leqq (1 - \frac{1}{4}a)t$ and also $\alpha_{J,0} \leqq (1 + \frac{1}{4}a)\tau_{J,0}$, then $\gamma_J(\tau_{J,0}) \leqq (1 - \frac{1}{16}a^2)t$. By (106–108): given $A_J(t)$, this fortunate conjunction has conditional P_i-probability at least $1 - a$, provided $0 < t \leqq d$. But f, the least t with $X(t) = j$, is $\gamma_J(\tau_{J,0})$ on $A_J \supset A_J(t)$, by (28). This completes the proof of (99). ★

The more familiar result that $\lim_{t \to 0} \gamma_J(t)/t = 1$ a.e. does not seem to be enough to establish (108), for the conditioning event is of small probability.

PROOF OF (90). Clear from (89). ★

PROOF OF (91). $Q_K = P'_K(0)$ exists by (37) and $(MC, 5.29)$. This has been used. Moreover, $Q_K(i, j)$ tends to a limit $Q_I(i, j)$, as argued above. Now use (89) and calculus to show that $P'(0, i, j)$ exists and is equal to $Q_I(i, j)$. ★

11. THE DISTRIBUTION OF X GIVEN X_J

This section, except for the basic theorem (109), will not be used elsewhere in the book, and arguments are only sketched. I recommend looking at $(MC,$ Sec. 4.5)† before tackling the present material. Continue in the setting of Section 5. Fix J with at least two elements and fix i in J. Recall that X_J visits $\xi_{J,0}, \xi_{J,1}, \ldots$ with holding times $\tau_{J,0}, \tau_{J,1}, \ldots$. Recall (65). Informally, $\alpha_{J,n}$ is the time X spends interior to the nth interval of constancy for X_J, and $\beta_{J,n}$ is the time X spends at the nth jump of X_J. For a moment, confine ω to $\{X(0) = i\}$. Then $\alpha_{J,0} + \beta_{J,1}$ is the least t with $X(t) \in J\setminus\{i\}$, by (28); and $\alpha_{J,0}$ is the sup of $s < \alpha_{J,0} + \beta_{J,1}$ with $X(s) = i$, by (25d); and $\tau_{J,0}$ is the time X spends in i from time 0 to time $\alpha_{J,0}$, by (28). Define a sequence of $\bar{I}$-valued processes $Y_0, Z_1, Y_1, Z_2, \ldots$ as in Figure 12:

$$Y_0(t) = X(t) \qquad\qquad \text{for } 0 \leq t < \alpha_{J,0}$$

$$Z_1(t) = X(t + \alpha_{J,0}) \qquad\qquad \text{for } 0 \leq t < \beta_{J,1}$$

$$Y_1(t) = X(t + \alpha_{J,0} + \beta_{J,1}) \qquad\qquad \text{for } 0 \leq t < \alpha_{J,1}$$

$$Z_2(t) = X(t + \alpha_{J,0} + \beta_{J,1} + \alpha_{J,1}) \quad \text{for } 0 \leq t < \beta_{J,2}$$

$$\vdots$$

$$i = 1 \quad J = \{1, 2\} \quad I = \{1, 2, 3, 4\}$$

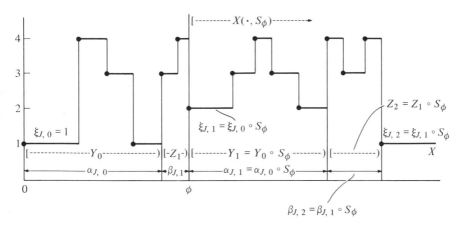

Figure 12.

†for motivation.

Thus, Y_n is X on the interval of X-times corresponding to X_J-times in the nth interval of constancy for X_J; and Z_n is X on the interval of X-times corresponding to the nth jump of X_J; both are shifted to the left so as to start at time 0. Of course, both are defined only on finite, random time intervals. Both have quasiregular sample functions with metrically perfect level sets, and visit φ on a set of times of Lebesgue measure 0. It is easy to check that $P_i\{Y_n(t) = \varphi\} = 0$ for any fixed $t \geq 0$. It is true that $P_i\{Z_n(t) = \varphi\} = 0$ for any fixed $t > 0$, although not necessarily for $t = 0$.

(109) Theorem. *Let $N = 0, 1, \ldots$. Given $\xi_{J,0}, \xi_{J,1}, \ldots, \xi_{J,N}$:*

(a) $(\tau_{J,0}, Y_0), Z_1, (\tau_{J,1}, Y_1), Z_2, \ldots, (\tau_{J,N}, Y_N), Z_{N+1}$ *are conditionally P_i-independent;*

(b) *for $1 \leq n \leq N$, the conditional P_i-distribution of $(\tau_{J,n}, Y_n)$ on $\{\xi_{J,n} = j\}$ coincides with the unconditional P_j-distribution of $(\tau_{J,0}, Y_0)$;*

(c) *for $1 \leq n \leq N$, the conditional P_i-distribution of Z_n on the set $\{\xi_{J,n-1} = j, \xi_{J,n} = k\}$ coincides with the P_j-distribution of Z_1 given $\{\xi_{J,1} = k\}$.*

NOTE. $(\tau_{J,0}, Y_0)$ is P_i-independent of $\{X(t + \alpha_{J,0}) : 0 \leq t < \infty\}$.

PROOF. Abbreviate $\alpha = \alpha_{J,0}$ and $\beta = \beta_{J,1}$. I will argue

(110) $\{X(t) : 0 \leq t < \alpha\}$ is P_i-independent of $\{X(\alpha + t) : 0 \leq t < \infty\}$.

Let

$$0 = t_0 < t_1 < \cdots < t_N < \infty \quad \text{and} \quad 0 = s_0 < s_1 < \cdots < s_M < \infty$$

$$i_0 = i, i_1, \ldots, i_N \quad \text{and} \quad j_0, j_1, \ldots, j_M \quad \text{be in } I$$

$$A = \{X(t_n) = i_n \text{ for } n = 0, \ldots, N \quad \text{and} \quad \alpha > t_N\}$$

$$B = \{X(\alpha + s_m) = j_m \text{ for } m = 0, \ldots, M\}.$$

Then (110) is equivalent to

(111) $P_i\{A \text{ and } B\} = P_i\{A\} \cdot P_i\{B\}.$

Let

$$\theta \text{ be the least } t \geq t_N \text{ with } X(t) = i$$

$$G = \{X(u) \notin J \setminus \{i\} \text{ for } 0 \leq u \leq \theta\}.$$

Check that θ is a Markov time, $X(\theta) = i$, and

$$A = \{X(t_n) = i_n \text{ for } n = 0, \ldots, N\} \cap G \in \mathcal{F}(\theta+).$$

Let W be the set of ω for which $X[\theta(\omega) + \cdot, \omega]$ is quasiregular, and let $S\omega$ be the retract of this function to R, the set of binary rationals in $[0, \infty)$.

As $(MC, 9.41a)$ implies, $P_i(W) = 1$. Look at Figure 13. Check that

$$\alpha = \theta + \alpha \circ S \quad \text{on } W \cap G \cap \{X(0) = i\},$$

so

$$A \cap B \cap W = A \cap S^{-1}B \cap W.$$

$$i = 1 \quad J = \{1, 2\} \quad I = \{1, 2, 3, 4\}$$

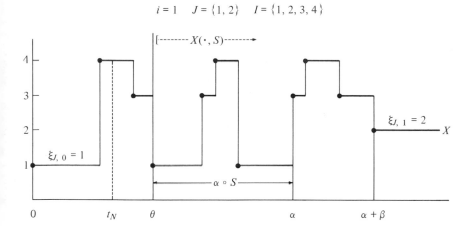

Figure 13.

Use strong Markov $(MC, 9.41c)$:

$$P_i\{A \cap B\} = P_i\{A \cap B \cap W\}$$
$$= P_i\{A \cap S^{-1}B \cap W\}$$
$$= P_i\{A \cap S^{-1}B\}$$
$$= P_i\{A\} \cdot P_i\{B\}.$$

This proves (111), and (110) follows from $(MC, 10.16)$, But $\tau_{J,0}$ is measurable on $Y_0 = \{X(t) : 0 \leqq t < \alpha\}$:

$$\tau_{J,0} = \text{Lebesgue } \{t : Y_0(t) = Y_0(0)\}.$$

And Z_1 is measurable on $X(\alpha + \cdot)$ modulo a P_i-null set: for Z_1 is $X(\alpha + \cdot)$ until first hitting $J\backslash\{i\}$, with P_i-probability 1. So (110) proves the case $N = 0$ of (109).

The general case follows inductively, using strong Markov on the time ϕ of first hitting $J \backslash \{X(0)\}$. In more detail, ϕ is Markov. Let W_ϕ be the set of ω for which $X[\phi(\omega) + \cdot, \omega]$ is quasiregular, and let $S_\phi \omega$ be the retract of this function to R, the set of binary rationals in $[0, \infty)$. As $(MC, 9.41a)$ implies, $P_i(W_\phi) = 1$. Look at Figure 12. Check that on $\{X(0) \in J\}$:
Figure 12. Check that on $\{X(0) \in J\}$:

$$\tau_{J,0}, \ Y_0, \text{ and } Z_1 \text{ are } \mathscr{F}(\phi+)\text{-measurable}$$

$$\xi_{J,1} = X(\phi) \in J \backslash \{X(0)\}.$$

Let $\sigma_0 = 0$. For $n \geq 1$, let

$$\sigma_n = \alpha_{J,0} + \cdots + \alpha_{J,n-1} + \beta_{J,1} + \cdots + \beta_{J,n}$$

$$= \gamma_J(\tau_{J,0} + \cdots + \tau_{J,n-1})$$

by (65). So $\sigma_1 = \phi$. On $\{X(0) \in J\}$, as (25d) and Figure 12 show,

$$\xi_{J,n} = X(\sigma_n)$$

$$\sigma_{n+1} \text{ is the least } t > \sigma_n \text{ with } X(t) \in J \backslash \{\xi_{J,n}\}$$

$$\tau_{J,n} = \text{Lebesgue } \{t : \sigma_n \leq t < \sigma_{n+1} \text{ and } X(t) = \xi_{J,n}\}$$

$$\sigma_{n+1} - \beta_{J,n+1} \text{ is the sup of } t < \sigma_{n+1} \text{ with } X(t) = \xi_{J,n}.$$

This generalizes (28). On $\{W_\phi \text{ and } X(0) \in J\}$: for $n = 1, 2, \ldots$

$$\xi_{J,n} = \xi_{J,n-1} \circ S_\phi \quad \text{and} \quad \tau_{J,n} = \tau_{J,n-1} \circ S_\phi$$

$$\alpha_{J,n} = \alpha_{J,n-1} \circ S_\phi \quad \text{and} \quad \beta_{J,n+1} = \beta_{J,n} \circ S_\phi$$

$$Y_n = Y_{n-1} \circ S_\phi \quad \text{and} \quad Z_{n+1} = Z_n \circ S_\phi.$$

See Figure 12. Fix $i \neq j$ in J. Strong Markov $(MC, 9.41d)$ implies: given $\{\xi_{J,1} = j\}$, the triple $(\tau_{J,0}, Y_0, Z_1)$ is conditionally P_i-independent of the infinite vector

$$\mathbf{V} = [(\xi_{J,n}, \tau_{J,n}, Y_n, Z_{n+1}) : n = 1, 2, \ldots];$$

and the conditional P_i-distribution of $\mathbf{V}$ coincides with the unconditional P_j-distribution of

$$[(\xi_{J,n}, \tau_{J,n}, Y_n, Z_{n+1}) : n = 0, 1, \ldots].$$

You do the induction.   ★

NOTE.   Theorem (109) includes (66), (81), and (84); it can in principle be derived from (81) and (84) by a passage to the limit.

The rest of this section is devoted to Y_0. Presumably, Z_1 can be treated in a similar way, although there is probably no analog to (121). Do not confine ω to $\{X(0) = i\}$, unless I specifically instruct you to do so. Let $(\alpha + \beta)$ be the least t with $X(t) \in J \setminus \{i\}$. Let α be the sup of $s < (\alpha + \beta)$ with $X(s) = i$. If no such s exists, let $\alpha = -\infty$. One drawback to this notation has to be faced: $(\alpha + \beta)$ is always a nonnegative number, and so is β when $\alpha > 0$. It is impossible for α to be 0. When $\alpha = -\infty$, the $+$ is purely symbolic, and β is undefined. Let

(112a) $\qquad \mu(t) = \text{Lebesgue } \{s : 0 \leq s \leq t \text{ and } X(s) = i\};$

(112b) $\qquad \tau = \mu(\alpha + \beta) = \mu(\alpha);$

(112c) $\qquad B(t) = \{\alpha + \beta > t\} = \{X(s) \notin J \setminus \{i\} \text{ for } 0 \leq s \leq t\};$

(112d) $\qquad B(t, k) = \{B(t) \text{ and } X(t) = k\};$

(112e) $\qquad h(j) = P_j\{\alpha > 0\} = P_j\{X \text{ hits } i \text{ before } J \setminus \{i\}\};$

(112f) $\qquad H = \{j : j = i \text{ or } j \in I \setminus J \text{ and } h(j) > 0\};$

(112g) $\qquad F(t, j, k) = \dfrac{h(k)}{h(j)} P_j\{B(t, k)\} \text{ for } j, k \in H;$

(112h) $\qquad \hat{F}(t, j, k) = \dfrac{h(k)}{h(j)} \displaystyle\int_{B(t,k)} e^{q\mu(t)} \, dP_j,$

where

$$q = q_J(i) \text{ and } j, k \in H.$$

By (25),

(113a) $\qquad \alpha = \alpha_{J,0} \text{ and } \beta = \beta_{J,1} \text{ and } \tau = \tau_{J,0} \text{ on } \{X(0) = i\}.$

As (25a) implies,

(113b) $\qquad\qquad\qquad \{\tau > t\} \subset B(t).$

Clearly,

(113c) $\qquad\qquad\qquad \tau = 0 \text{ iff } \alpha = -\infty$

(113d) $\qquad\qquad\qquad h(i) = 1.$

(114) Theorem.   *Relative to P_j, the fragment $\{X(t) : 0 \leq t < \alpha\}$ is Markov with stationary, standard transitions F_i and state space H, when $j \in H$.*

PROOF.   Let $0 = t_0 < t_1 < \cdots < t_N < \infty$ and let $j_0 = j, j_1, \ldots, j_N$ be in H. Let A be the event $\{X(t_n) = j_n \text{ for } n = 1, \ldots, N \text{ and } \alpha > t_N\}$.

My problem is to compute $P_j(A)$. Abbreviate $t = t_1$ and $k = j_1$. Use the notation and result of the Markov property $(MC, 9.31)$. Confine ω to W_t, which has P_j-probability 1. Then

$$\{\alpha > t + s\} = B(t) \cap \{\alpha \circ T_t > s\} \quad \text{for any } s \geq 0,$$

and

$$A = B(t, k) \cap T_t^{-1}\hat{A},$$

where

$$\hat{A} = \{X(t_n - t) = j_n \text{ for } n = 2, \ldots, N \text{ and } \alpha > t_N - t\}.$$

Hence,

$$P_j(A) = P_j\{B(t, k)\} \cdot P_k(\hat{A}).$$

Remember that

$$P_j\{X(0) = j \text{ and } \alpha > 0\} = P_j\{\alpha > 0\} = h(j).$$

Induct:

$$P_j(A) = h(j) \prod_{n=0}^{N-1} F(t_{n+1} - t_n, j_n, j_{n+1}).$$

As in $(MC, 5.4)$, this makes F a substochastic semigroup, and also makes $\{X(t): 0 \leq t < \alpha\}$ Markov with transitions F, and F is standard by (17). ★

The next result leads to a sharpening of (114). The statement uses (112). Let $0 = t_0 < t_1 < \cdots < t_N < s < \infty$. Let $j_0 = j, j_1, \ldots, j_N \in H$.

(115) Lemma. *The P_j-probability that $X(t_n) = j_n$ for $n = 1, \ldots, N$ and $\tau \geq s$ is*

$$h(j)e^{-qs} \prod_{n=0}^{N-1} \hat{F}(t_{n+1} - t_n, j_n, j_{n+1}).$$

PROOF. Begin with the case $N = 1$. Abbreviate $t = t_1$ and $k = j_1$. My problem is to evaluate $P_j(A)$, where A is the event $\{X(t) = k \text{ and } \tau \geq s\}$. Use the notation and result of the Markov property $(MC, 9.31)$. Confine ω to W_t, which has P_j-probability 1. Review (112a–c). Put $J = \{i\}$ and $s = \alpha + \beta - t$ in (49a) to get

$$(116) \qquad\qquad \tau = \mu(t) + \tau(T_t) \quad \text{on } B(t);$$

because $\alpha + \beta - t$ is the time $X(\cdot, T_t)$ hits $J \setminus \{i\}$. Review (112d). By (116):

$$A = \{B(t, k) \text{ and } \tau(T_t) \geq s - \mu(t)\}.$$

Let

$$f(x) = P_k\{\tau \geq x\}.$$

Remember $\mu(t) \leq t < s$. I say

(117) $$P_j\{A\} = \int_{B(t,k)} f[s - \mu(t)]\, dP_j.$$

To prove (117), let

$$F(\omega, \omega') = 1 \quad \text{if } \omega \in B(t, k) \text{ and } \tau(\omega') \geq s - \mu(t, \omega)$$
$$= 0 \quad \text{elsewhere.}$$

Let $\mathscr{F}$ be the product σ-field in Ω relativized to Ω_∞. Both $B(t, k)$ and $\mu(t)$ are $\mathscr{F}(t)$-measurable. So F is $\mathscr{F}(t) \times \mathscr{F}$-measurable on $\Omega_\infty \times \Omega_\infty$. Recognize $1_A = F(\cdot, T_t)$ and use $(MC, 9.31e)$:

$$P_j\{A\} = \int F(\omega, T_t\omega)\, P_j(d\omega) = \int F^*(\omega)\, P_j(d\omega).$$

If $\omega \in B(t, k) \subset \{X(t) = k\}$, then

$$F^*(\omega) = \int F(\omega, \omega')\, P_{X(t,\omega)}(d\omega')$$
$$= P_k\{\omega' : \tau(\omega') \geq s - \mu(t, \omega)\}$$
$$= f[s - \mu(t, \omega)].$$

If $\omega \notin B(t, k)$, then $F^*(\omega) = 0$. This proves (117). Review (112e), and remember $q = q_J(i)$. If $x > 0$, then

(118) $$f(x) = P_k\{\tau \geq x\}$$
$$= h(k)P_i\{\tau \geq x\}$$
$$= h(k)e^{-qx}:$$

the first equality comes from strong Markov on P_k and the time of first hitting i; and the second from (44). Consequently,

(119) $$P_j(A) = e^{-qs} \int_{B(t,k)} e^{q\mu(t)}\, dP_j.$$

Now try the general case, inductively. Let A be the event that $X(t_n) = j_n$ for $n = 1, \ldots, N$. Again abbreviate $t = t_1$ and $k = j_1$. Let $\hat{A}$ be the event that $X(t_n - t) = j_n$ for $n = 2, \ldots, N$. Again use $(MC, 9.31)$ and confine ω to W_t. As (116) implies,

$$\{A \text{ and } \tau \geq s\} = B(t, k) \cap C$$

where

$$C = \{T_t^{-1}\hat{A} \text{ and } \tau(T_t) \geq s - \mu(t)\}.$$

Let

$$f(x) = P_k\{\hat{A} \text{ and } \tau \geq x\}.$$

I say

(120) $$P_j\{A \text{ and } \tau \geq s\} = \int_{B(t,k)} f[s - \mu(t)] \, dP_j.$$

This can be argued like (117). Now $\mu(t) \leq t$, so $s - \mu(t) > t_N - t$. By inductive assumption,

$$f(x) = h(k)e^{-qx} \Pi_{n=1}^{N-1} \hat{F}(t_{n+1} - t_n, j_n, j_{n+1}) \quad \text{for } x > t_N - t.$$

Reorganizing,

$$P_j\{A \text{ and } \tau \geq s\} = h(k)e^{-qs} \int_{B(t,k)} e^{q\mu(t)} \, dP_j \cdot \Pi_{n=1}^{N-1} \hat{F}(t_{n+1} - t_n, j_n, j_{n+1})$$

$$= h(j) \Pi_{n=0}^{N-1} \hat{F}(t_{n+1} - t_n, j_n, j_{n+1}). \qquad \bigstar$$

Suppose Y is a jointly measurable process, and t a nonnegative real number, or even a random variable. Let $\lambda(Y, t)$ be the least s such that

$$\text{Lebesgue } \{u : 0 \leq u \leq s \text{ and } Y(u) = i\} = t.$$

In particular, $\lambda(X, \tau_{J,0}) = \alpha_{J,0}$ on $\{X(0) = i\}$. Clearly, $\lambda(Y, t) \geq t$. Review (112–113).

(121) Theorem. *Fix $j \in H$. On a suitable probability triple, there is a random variable T and a process $\{Y(t) : 0 \leq t \leq \infty\}$ with the following properties:*

(a) *The joint distribution of T and $\{Y(t) : 0 \leq t < \lambda(Y, T)\}$ coincides with the joint P_j-distribution of τ and $\{X(t) : 0 \leq t < \alpha\}$.*

(b) *T and Y are independent.*

(c) *$Y(0) = j$.*

(d) *Y has quasiregular sample functions, with values in $H \cup \{\varphi\}$.*

(e) *$\{t : Y(t) = j\}$ is metrically perfect and has infinite Lebesgue measure for all j in H.*

(f) *$\{t : Y(t) = \varphi\}$ has Lebesgue measure 0.*

(g) *Prob $\{Y(t) = \varphi\} = 0$.*

Furthermore, $\hat{F}$ is a standard stochastic semigroup, and Y is Markov with stationary transitions $\hat{F}$.

PROOF. Properties (a) and (b) follow as in (56). Indeed, $P_j\{\cdot | \tau \geq s\}$ retracted to $\mathscr{F}[\gamma_j(t) +]$ does not depend on $s > t$, by strong Markov. This enables you to construct T and $\{Y(r) : 0 \leq r < \infty$ and r is a binary rational$\}$

so as to satisfy (a) and (b) with t confined to the binary rationals. Moreover, $Y(0) = j$ because $X(0) = j$, after discarding two null sets, giving (c). Next, X is quasiregular. Using (a), (b) and the unboundedness of T, one can discard a null set so that Y will be quasiregular on the binary rationals. Then Y can be extended to $[0, \infty)$ so as to satisfy (a), (b), (c), (d). Now (a), (b) and the unboundedness of T give (e), (f), (g) from the corresponding properties of X, again after deleting a null set.

For the last assertion, let $0 = t_0 < t_1 < \cdots < t_N$ and let $j_0 = j_1, \ldots, j_N$ be in H. Use (115), (118), and (121a–b):

$$\text{Prob}\,\{Y(t_n) = j_n \quad \text{for } n = 0, \ldots, N\} = \prod_{n=0}^{N} \hat{F}(t_{n+1} - t_n, j_n, j_{n+1}).$$

Use (121d, f, g) and (MC, 5.4) to make $\hat{F}$ a stochastic semigroup and Y Markov with stationary transitions $\hat{F}$. Then use (121d) and (17) to make $\hat{F}$ standard.   ★

NOTE. $P_i\{\tau = \tau_{J,0}\} = P_i\{\alpha = \alpha_{J,0}\} = 1$, by (113b).

WARNING. The distribution of Y is usually different from the distribution of X restricted to H. Informally, Y is X conditioned on spending an infinite amount of time in i before hitting $J \setminus \{i\}$. In particular, X is conditioned on reaching i before hitting $J \setminus \{i\}$.

Of course, (121) and (64*) prove (56). And (121) includes (114). In current terminology, the fragment $\{Y(t):0 \leq t < \lambda(Y, T)\}$ is obtained from Y by killing Y at the rate $q = q_J(i)$ along the set $\{t: Y(t) = i\}$.

Another proof of (121)

It is possible to prove (121) by a more daring and instructive method. To simplify the writing, adopt the notation of Section 9. Number the states so $i = 1$ and $J = I_N$ and $H = \{1, N + 1, N + 2, \ldots\}$. This assumes H is infinite: the finite case is easier. Write X_N for X_{I_N}, and Q_N for Q_{I_N}, and $\tau_{N,0}$ for $\tau_{I_N,0}$. Recall that $q_N(j) = -Q_N(j, j)$ and $\Gamma_N(j, k) = Q_N(j, k)/q_N(j)$ for $j \neq k$. On a convenient probability triple, construct inductively a sequence Y_N, $Y_{N+1}, \ldots$ of stochastic processes with time parameter running through $[0, \infty)$. Namely, Y_N is identically equal to 1. The process Y_{N+m+1} has right continuous step functions for sample functions, with values in the set $\{1, N + 1, \ldots, N + m + 1\}$. It is obtained from Y_{N+m} by cutting Y_{N+m} and inserting $(N + m + 1)$-intervals. Given Y_{N+m} and the locations of the cuts, the lengths of the inserted intervals are independent and exponentially distributed, with common parameter $q_{N+m+1}(N + m + 1)$. There are two kinds of cuts: the first kind appears at a jump of Y_{N+m}, and the second kind appears interior to an interval of constancy for Y_{N+m}. Cuts of the first kind appear independently from jump to jump, and independently of cuts of the second kind. Locations of cuts of the second kind are independent from interval to interval. At a jump from j to k, the probability of a cut not appearing is

$$Q_{N+m+1}(j,k)/Q_{N+m}(j,k).$$

Within a particular j-interval, the location of cuts has a Poisson distribution, with parameter

$$q_{N+m+1}(j) - q_{N+m}(j).$$

On the same triple, construct a random variable T independent of the Y's, and exponentially distributed with parameter $q = q_N(1)$. A formal description of all this is by now easy and is omitted.

(122) Lemma.   *The joint distribution of T and the fragment*

$$\{Y_{N+m}(t):0 \leqq t < \lambda(Y_{N+m}, T)\}$$

coincides with the joint P_1-distribution of $\tau_{N,0}$ and

$$\{X_{N+m}(t):0 \leqq t < \lambda(X_{N+m}, \tau_{N,0})\}.$$

PROOF.   Use (85).   ★

(123) Lemma. Y_{N+m} *is Markov, with stationary standard transitions.*

PROOF.   This is trivial for $m = 0$. Suppose Y_{N+m} is Markov, with generator $\hat{Q}_{N+m}$. In view of (86), I have to guess a generator $\hat{Q}_{N+m+1}$ on $\{1, N+1, \ldots, N+m+1\}$ whose restriction to $\{1, N+1, \ldots, N+m\}$ is $\hat{Q}_{N+m}$, and which gives the proper instructions for constructing Y_{N+m+1} from Y_{N+m}. Then Y_{N+m+1} will be Markov with generator $\hat{Q}_{N+m+1}$. The guess is motivated by Section 4.5 of MC and lemma (71). Let

(124) $\hat{q}_{N+m+1}(1) = q_{N+m+1}(1) - q_N(1)$

and

(125) $\hat{q}_{N+m+1}(j) = q_{N+m+1}(j)$ for $j = N+1, \ldots, N+m+1.$

For $j = 1, N+1, \ldots, N+m+1$, let $g_{N+m+1}(j)$ be the probability that a discrete time Markov chain with starting state j and stationary transitions Γ_{N+m+1} reaches 1 in positive time before hitting $\{2, \ldots, N\}$. Verify that $g_{N+m+1}(j)$ does not depend on m, provided $N+m+1 \geqq j > N$. Call the common value $g(j)$. Clearly, $g(j)$ is the P_j-probability that X reaches 1 before hitting $\{2, \ldots, N\}$. So $g(j) = h(j) > 0$. The dependence of $g_{N+m+1}(1)$ on m is simple. As for (73a), or using (87f),

(126) $q_N(1) = [1 - g_{N+m+1}(1)]q_{N+m+1}(1).$

Without real loss, suppose $g_{N+m+1}(1) > 0$. For $j \neq k$ in $\{N + 1, \ldots, N + m\}$, let

(127) $\hat{\Gamma}_{N+m+1}(j, k) = \dfrac{g(k)}{g(j)}\Gamma_{N+m+1}(j, k).$

For $j = N + 1, \ldots, N + m + 1$, let

(128) $\hat{\Gamma}_{N+m+1}(j, 1) = \dfrac{1}{g(j)}\Gamma_{N+m+1}(j, 1).$

For $j = N + 1, \ldots, N + m + 1$, let

(129) $\hat{\Gamma}_{N+m+1}(1, j) = \dfrac{g(j)}{g_{N+m+1}(1)}\Gamma_{N+m+1}(1, j).$

For $j \neq k$ in $\{1, N + 1, \ldots, N + m + 1\}$, let

(130) $\hat{Q}_{N+m+1}(j, k) = \hat{q}_{N+m+1}(j)\hat{\Gamma}_{N+m+1}(j, k).$

For j in $\{1, N + 1, \ldots, N + m + 1\}$, let

(131) $\hat{Q}_{N+m+1}(j, j) = -\hat{q}_{N+m+1}(j).$

This completes the construction of $\hat{Q}$.

For the verification, extend $\hat{\Gamma}$ to vanish on the diagonal, and use strong Markov to check that the extended matrix is stochastic. Use $(MC, 5.29)$ and (129–130) to see that $\hat{Q}$ is a generator. Check that (72a) holds with hats on. Proceed to check that the restriction of $\hat{Q}_{N+m+1}$ to $\{1, N + 1, \ldots, N + m\}$ gives $\hat{Q}_{N+m}$. Remember π from (72b). Begin by writing down (73b) for the Q-sequence, with $N + m$ for n and $i \neq j$ in $\{N + 1, \ldots, N + m\}$. Multiply across by $g(j)/g(i)$ and use (127) to get (73b) for the $\hat{Q}$-sequence, because $\pi_{N+m+1}(i)$ is the same for Q_{N+m+1} as for $\hat{Q}_{N+m+1}$, again by (127). The same program succeeds for $i \in \{N + 1, \ldots, N + m\}$ and $j = 1$, multiplying across this time by $1/g(i)$ and using (128). The condition (73a) is easy for $i \geq N$, using (125). Turn to row 1. There, save trouble by checking from (124, 126) that

(132a) $\hat{q}_{N+m+1}(1) = g_{N+m+1}(1)q_{N+m+1}(1),$

so (130) makes

(132b) $\hat{Q}_{N+m+1}(1, j) = Q_{N+m+1}(1, j)g(j)$ for $j = N + 1, \ldots, N + m + 1$.

Write down (73c) for the Q-sequence with $i = 1$ and $j \geq N + 1$ and $N + m$ for n; multiply across by $g(j)$ and get (73c) for the $\hat{Q}$-sequence. There remains condition (73a) for $i = 1$, which is a bore. Namely, $\hat{\pi}_{N+m+1}(1)$ is

$$1 - \hat{\Gamma}_{N+m+1}(1, N + m + 1)\hat{\Gamma}_{N+m+1}(N + m + 1, 1)$$

which equals

$$1 - [\Gamma_{N+m+1}(1, N + m + 1)\Gamma_{N+m+1}(N + m + 1, 1)/g_{N+m+1}(1)]$$

by (72b) and (128–129). So

(133) $$g_{N+m+1}(1)\hat{\pi}_{N+m+1}(1) = g_{N+m+1}(1) - 1 + \pi_{N+m+1}(1).$$

To push on, $\hat{\pi}_{N+m+1}(1)\hat{q}_{N+m+1}(1)$ is

$$\hat{\pi}_{N+m+1}(1)g_{N+m+1}(1)q_{N+m+1}(1) \tag{132a}$$

$$= g_{N+m+1}(1)q_{N+m+1}(1) - q_{N+m+1}(1) + \pi_{N+m+1}(1)q_{N+m+1}(1) \tag{133}$$

$$= \hat{q}_{N+m+1}(1) - q_{N+m+1}(1) + q_{N+m}(1) \tag{132a, 73a}$$

$$= q_{N+m}(1) - q_N(1) \tag{124}$$

$$= \hat{q}_{N+m}(1). \tag{124}$$

You should check that the instructions for making Y_{N+m+1} from Y_{N+n} are right: namely,

$$\hat{q}_{N+m+1}(i) - \hat{q}_{N+m}(i) = q_{N+m+1}(i) - q_{N+m}(i) \text{ for } i \text{ in } \{1, N + 1, \ldots, N + m\};$$

$$\frac{\hat{Q}_{N+m+1}(i, j)}{\hat{Q}_{N+m}(i, j)} = \frac{Q_{N+m+1}(i, j)}{Q_{N+m}(i, j)} \qquad \text{for } i \neq j \text{ in } \{1, N + 1, \ldots, N + m\}.$$

This is algebra, of the same kind I've been doing.

To finish the second proof of (121), argue that $\{Y_m\}$ determines a limit Y, such that Y_{N+m} is the restriction of Y to $\{1, N + 1, \cdots, N + m\}$. This can be deduced from (3.5) below, using (122) and (123). Continue by verifying that Y has the required properties, using the same facts and (34). This passage to the limit is delicate.   ★

Displaying X on the X_N scale

In thinking about these constructions, there is one idea I found very helpful but too cumbersome when described formally: the possibility of displaying X on the X_N-scale. Without real loss, suppose $I = \{1, 2, 3\}$ and $N = 1$. Then X_2 is obtained from X_1 by inserting 2-intervals, and $X = X_3$ is obtained from X_2 by inserting 3-intervals. For the last stage, what counts is the relative position of the 3-cuts inside the 2-intervals, and the position of the 3-cuts inside the 1-intervals referred to the X_1-scale. The whole procedure can then be represented as in Figure 14.

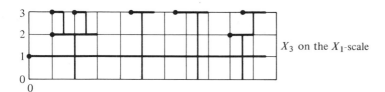

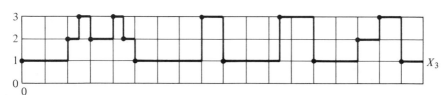

Figure 14.

2

RESTRICTING THE RANGE: APPLICATIONS

1. FOREWORD

Some general theorems on Markov chains are proved in Sections 2 through 6, using the theory developed in Chapter 1. In Section 7, there are hints for dealing with the transient case. A summary can be found at the beginning of Chapter 1. The sections of this chapter are almost independent of one another. For Sections 2 through 6, continue in the setting of Section 1.5. Namely, I is a finite or countably infinite set, with the discrete topology; and $\bar{I} = I$ for finite I, while $\bar{I} = I \cup \{\varphi\}$ is the one-point compactification of I for infinite I. There is a standard stochastic semigroup P on I, for which each i is recurrent and communicates with each j. For a discussion, see Section 1.4. The process X on the probability triple (Ω_∞, P_i) is Markov with stationary transitions P, starting state i, and smooth sample functions. Namely, the sample functions are quasiregular and have metrically perfect level sets with infinite Lebesgue measure. For finite $J \subset I$, the process X_J is X watched only when in J. This process has J-valued right continuous sample functions, which visit each state on a set of times of infinite Lebesgue measure. Relative to P_i, the process X_J is Markov with stationary transitions P_J, and generator $Q_J = P'_J(0)$. For a discussion, see Section 1.6. Recall that X_J visits states $\xi_{J,0}, \xi_{J,1}, \ldots$ with holding times $\tau_{J,0}, \tau_{J,1}, \ldots$. Recall that $\mu_J(t)$ is the time on the X_J-scale corresponding to time t on the X-scale, while $\gamma_J(t)$ is the largest time on the X-scale corresponding to time t on the X_J-scale.

2. THE GENERATOR

Fix $i \in I$. Say X *pseudo-jumps* from i to j in I at time $\sigma > 0$ iff $X(\sigma) = j$ while $q\text{-}\lim_{t \downarrow 0} X(\sigma - t) = i$; that is, there are binary rational $r > \sigma$ but

64

arbitrarily close to it with $X(r) = j$, while there are binary rational $s < \sigma$ but arbitrarily close to it with $X(s) = i$. By quasiregularity, the times of the pseudo-jumps from i to j are isolated; this was observed in $(MC, 9.25)$. Let $0 < \sigma_0(j) < \sigma_1(j) < \cdots$ be the times of the pseudo-jumps from i to j, numbered from 0. If there are n or fewer of these pseudo-jumps, put $\sigma_n(j) = \infty$. Let

$$\theta_n(j) = \mu_{\{i\}}[\sigma_n(j)]$$

$$= \text{Lebesgue } \{s: 0 \leq s \leq \sigma_n(j) \text{ and } X(s) = i\}$$

be the time on the $X_{\{i\}}$-scale corresponding to the nth pseudo-jump from i to j. Again, $\theta_n(j) = \infty$ if there are n or fewer of these pseudo-jumps. Recall that $Q = P'(0)$ and $q(i) = -Q(i, i)$ and $\Gamma(i, j) = Q(i, j)/q(i)$. Assumption (1.24) is in force.

(1) Theorem. $Q(i, j) = 0$ iff there are no pseudo-jumps from i to j, with P_i-probability 1.

To avoid exceptional cases, a Poisson process with parameter 0 consists with probability 1 of the single point ∞. Then (1) is included in the main result

(2) Theorem. With respect to P_i, the process $\{\theta_n(j): n = 0, 1, \ldots\}$ is Poisson with parameter $Q(i, j)$: and these processes are independent as j varies over $I \setminus \{i\}$.

PROOFS. *Finite I.* Suppose first that I is finite. Then there is no difference between pseudo-jumps and jumps. As in Figure 1, suppose X visits $\xi_0, \xi_1, \ldots$ with holding times $\tau_0, \tau_1, \ldots$. Confine ω to $\{X(0) = i\}$, so $\xi_0 = i$. Theorem (1) is immediate from $(MC, 5.45)$. I will argue (2) from (1.71). Let the process $\mathcal{P} = (T_0, T_1, \ldots)$ have the holding times of X in i as its interarrival times. Namely, let $0 = n(0) < n(1) < \ldots$ be the indices n with $\xi_n = i$, and let

$$T_0 = \tau_{n(0)}, \quad \text{and } T_1 = T_0 + \tau_{n(1)}, \quad \text{and } T_2 = T_1 + \tau_{n(2)}, \cdots.$$

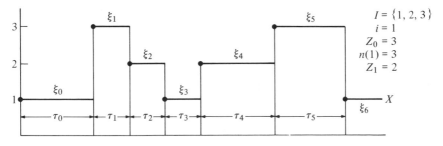

Figure 1.

Let the finite set F be $I \setminus \{i\}$. Let $Z_m = \xi_{n(m)+1}$. That is, Z_m is the state X jumps to on leaving the mth i-interval, with $m = 0, 1, \ldots$. See Figure 1. With respect to P_i: the process $\mathscr{P}$ is Poisson with parameter $q(i)$; the random variables Z are independent with common distribution $P_i\{Z_m = j\} = \Gamma(i, j)$: and Z is independent of $\mathscr{P}$. Indeed, let M and N be nonnegative integers. Let $i_1, \ldots, i_N$ be any I-sequence with at least M visits to i. Let $t_0, \ldots, t_M$ be nonnegative numbers. Let $i_0 = i$ and $T_{-1} = 0$. By $(MC, 5.45)$,

$$P_i\{T_m - T_{m-1} > t_m \text{ for } m = 0, \ldots, M \text{ and } \xi_n = i_n \text{ for } n = 0, \ldots, N\}$$
$$= e^{-t} P_i\{\xi_n = i_n \text{ for } n = 0, \ldots, N\},$$

where

$$t = q(i) \, \Sigma_{m=0}^{M} \, t_m.$$

Sum over all possible sequences $i_0, i_1, \ldots, i_N$ which have exactly $M + 1$ terms equal to i, and $i_0 = i_N = i$; even N is allowed to vary. The sets

$$\{\xi_n = i_n \text{ for } n = 0, \ldots, N\}$$

are disjoint, and their union is $\{X(0) = i\}$. So the sum of the left sides is

$$P_i\{T_m - T_{m-1} \overset{.}{>} t_m \text{ for } m = 0, \ldots, M\},$$

and the sum of the right sides is e^{-t}. That is,

$$P_i\{T_m - T_{m-1} > t_m \text{ for } m = 0, \ldots, M\} = e^{-t}.$$

This and $(MC, 10.16)$ prove: $\mathscr{P}$ is Poisson with parameter $q(i)$, and is independent even of $\{\xi_n\}$. Now $\{\xi_n\}$ is Markov with transitions Γ starting from i, and $\{Z_m\}$ is a function of $\{\xi_n\}$. Use strong Markov $(MC, 1.21\}$ on $\{\xi_n\}$ to get the distribution of $\{Z_m\}$. The process $\mathscr{P}_j$ of (1.71), consisting of those T_m with $Z_m = j$, is clearly the process $\{\theta_n(j): n = 0, 1, \ldots \}$. These processes are consequently independent and Poisson, with parameter $q(i)\Gamma(i, j) = Q(i, j)$. This completes the proof for finite I.

   Infinite I.   Fix i, confine ω to $\{X(0) = i\}$, and compute relative to P_i. Let K be a finite subset of I which contains i. Using (1.37), apply the results for finite I to the process X_K. Namely, let $\sigma_{K,n}(j)$ be the X_K-time of the nth jump from i to j in X_K. Let $\theta_{K,n}(j)$ be the time X_K spends in i up to $\sigma_{K,n}(j)$:

$$\theta_{K,n}(j) = \text{Lebesgue } \{s: 0 \leq s \leq \sigma_{K,n}(j) \text{ and } X_K(s) = i\}.$$

Then $\{\theta_{K,n}(j): n = 0, 1, \ldots\}$ is a Poisson process with parameter $Q_K(i, j)$, and these processes are independent as j varies over $K \setminus \{i\}$. Now

$$Q_K(i, j) \downarrow Q(i, j) \quad \text{as } K \uparrow I$$

by (1.90). The only thing left to check is

(3) $\theta_{K,n}(j) \to \theta_n(j)$ as $K \uparrow I$ for each n and j.

   This will be obvious on reflection, but here is a formal proof. Only the
case $n = 0$ will be argued, the rest being similar. Fix $j \neq i$ in I. Let $J = \{i, j\}$
and suppose $K \supset J$. It is helpful to think of X_J, as in Figure 2. Abbreviate

$$S_m = \sigma_{J,m}(j) = \Sigma_{n=0}^{2m} \tau_{J,n}$$

and

$$T_m = \theta_{J,m}(j) = \Sigma_{n=0}^{m} \tau_{J,2n}.$$

Let $S_\infty = \lim_{m \to \infty} S_m = \infty$ and $T_\infty = \lim_{m \to \infty} T_m = \infty$. Let $m(I)$ be the
least $m = 0, 1, \ldots$ if any with $\gamma_J(S_m-) = \gamma_J(S_m)$, and $m(I) = \infty$ if no such m
exists. Let $\mu_J(\infty) = \gamma_J(\infty) = \infty$.

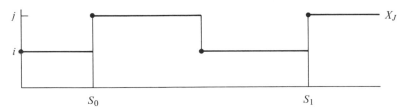

Figure 2.

   The main step in proving (3) is

(4) $\sigma_0(j) = \gamma_J(S_{m(I)})$.

To see this, remember (1.25c) that μ_J is a continuous, strictly increasing map
of $\{t : X(t) \in J\}$ onto $[0, \infty)$. If $\sigma_0(j) = \infty$, then $\mu_J[\sigma_0(j)] = \infty$. If $\sigma_0(j) < \infty$,
then $\mu_J[\sigma_0(j)]$ is the J-time of a jump from i to j in X_J, namely an S_m for some
m. Thus, $\sigma_0(j) = \gamma_J(S_m)$ for some m. Next, γ_J is a right continuous strictly
increasing map of $[0, \infty)$ onto $\{t : X(t) \in J\}$, by (1.25d). If $\gamma_J(S_m-) < \gamma_J(S_m)$,
then $\gamma_J(S_m)$ cannot be the time of a pseudo-jump from i to j in X. On the
other hand, if $\gamma_J(S_m-) = \gamma_J(S_m)$, then both give the time of a pseudo-jump
from i to j in X. Indeed,

$$X[\gamma_J(S_m + t)] = X_J(S_m + t);$$

let $t \downarrow 0$ and $t \uparrow 0$. Consequently, the set of times $\sigma_0(j), \sigma_1(j), \ldots$ coincides with
the set of times $\gamma_J(S_m)$ with $\gamma_J(S_m-) = \gamma_J(S_m)$ and $m = 0, 1, \ldots$. This completes the proof of (4).
   Now I claim

(5) $\theta_0(j) = T_{m(I)}$.

Indeed

$$\theta_0(j) = \mu_{\{i\}}[\sigma_0(j)] \qquad \text{(definition)}$$

$$= \mu_{\{i\}}[\gamma_J(S_{m(I)})] \qquad \text{(4)}$$

$$= \mu_{\{i\},J}[S_{m(I)}] \qquad \text{(1.32a)}$$

$$= T_{m(I)} \qquad \text{(definition)}.$$

Let $m(K)$ be the least m with $\gamma_{J,K}(S_m -) = \gamma_{J,K}(S_m)$, and $m(K) = \infty$ if no such m exists. The definition of $\gamma_{J,K}$ is in Section 1.5. Use (1.35) to get these analogs of (4) and (5):

(6) $$\sigma_{K,0}(j) = \gamma_{J,K}(S_{m(K)})$$

(7) $$\theta_{K,0}(j) = T_{m(K)}.$$

Now use (1.34c) to check that $m(K) = m(I)$ for all sufficiently large K in case $m(I) < \infty$, while $m(K) = \infty$ in case $m(I) = \infty$. Actually, $m(K) \geq m(I)$ in all cases. This completes the proof of (3), in case $n = 0$. In fact, I proved:

$$\theta_{K,0}(j) \geq \theta_0(j) \quad \text{for all } K;$$

$$\theta_{K,0}(j) = \theta_0(j) \quad \text{for all large } K. \qquad \qquad \bigstar$$

3. A THEOREM OF LÉVY

The object of this section is to prove the following dichotomy of Lévy (1952, 1958).

(8) Theorem. *For each pair $i \neq j$ in I, there are only two possibilities: either $P(t, i, j) = 0$ for all $t \geq 0$, or $P(t, i, j) > 0$ for all $t > 0$.*

PROOF. This is true even without assumption (1.24). For a moment, drop this assumption. Fix $i \neq j$ in I. As $(MC, 5.7)$ implies, $P(t, s, s) > 0$ for all $s \geq 0$. Suppose $P(t, i, j) > 0$ for some $t > 0$. Then

$$P(t + s, i, j) \geq P(t, i, j)P(s, j, j) > 0.$$

The difficult assertion in (8) is:

(9) $$P(t, i, j) > 0 \text{ implies } P(u, i, j) > 0 \text{ for some } u \leq \tfrac{1}{2}t.$$

It is enough to do the case $t = 2$. Introduce f, the least s with $X(s) = j$. Now $\{f \leq 2\} \supset \{X(2) = j\}$, so $P(2, i, j) > 0$ implies

(10) $$P_i\{f \leq 2\} > 0.$$

By quasiregularity,

$$\{f \leq 1\} \subset \{X(r) = j \text{ for some binary rational } r \text{ in } [0, 1]\}.$$

By countable additivity, it is enough to deduce

(11) $$P_i\{f \leq 1\} > 0$$

from (10). The reasoning is perfectly general, but to avoid mess, assume i is recurrent. Reduce I to the set of k which communicate with i; this contains j. Then (1.24) is back in force.

Let J be a (large) finite subset of I, which contains i and j. Recall that f is the least t with $X(t) = j$, and introduce

$$\varepsilon_J = \text{Lebesgue } \{t : 0 \leq t \leq f \text{ and } X(t) \notin J\}.$$

As (1.33a) implies, $\varepsilon_J \downarrow 0$ as $J \uparrow I$. Use (10) and countable additivity to fix $J \supset \{i, j\}$ so large that

(12) $$P_i\{\varepsilon_J \leq \tfrac{1}{2} \text{ and } f \leq 2\} > 0.$$

Consider X_J. Let f_J be the least t with $X_J(t) = j$. Since $f_J \leq f$, relation (12) implies

(13) $$P_i\{\varepsilon_J \leq \tfrac{1}{2} \text{ and } f_J \leq 2\} > 0.$$

The idea for the rest of the proof is to divide the X_J-time scale by 4. This brings f_J down from 2 to $\tfrac{1}{2}$, tends to reduce ε_J, and does not change X_J-sets of positive probability to zero probability. The coup de grâce is provided by (15) below.

Recall that X_J visits states $\xi_{J,0}, \xi_{J,1}, \ldots$ with holding times $\tau_{J,0}, \tau_{J,1}, \ldots$. Use countable additivity on (13) to find a nonnegative integer N and states $i_1, i_2, \ldots, i_N$ in $J \setminus \{j\}$, such that

$$P_i\{A \text{ and } \varepsilon_J \leq \tfrac{1}{2} \text{ and } f_J \leq 2\} > 0,$$

where A is $\{\xi_{J,0} = i, \xi_{J,1} = j\}$ when $N = 0$, while for $N \geq 1$,

$$A = \{\xi_{J,0} = i, \xi_{J,1} = i_1, \ldots, \xi_{J,N} = i_N, \xi_{J,N+1} = j\}.$$

See Figure 3 for $N = 1$. Thus

(14) $$P_i\{A\} > 0 \quad \text{and} \quad P_i\{\varepsilon_J \leq \tfrac{1}{2} \text{ and } f_J \leq 2 | A\} > 0.$$

From Section 1.8, recall that $\alpha_{J,n}$ is the time X spends interior to the nth interval of constancy for X_J, and $\beta_{J,n}$ is the time X spends at the nth jump of X_J. On A:

(15) $$f = f_J + \varepsilon_J;$$

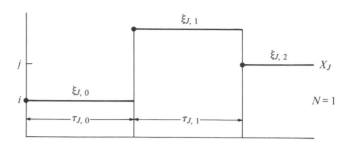

Figure 3.

(16) $$f_J = \Sigma_{n=0}^{N} \tau_{J,n};$$

(17) $$\varepsilon_J = \Sigma_{n=1}^{N+1} \beta_{J,n} + \Sigma_{n=0}^{N} (\alpha_{J,n} - \tau_{J,n}).$$

This is intuitively clear, and (16) is even formally clear. To get (15), use (1.25) to check $f = \gamma_J(f_J)$: so $f_J = \mu_J(f)$ and

$$\varepsilon_J = f - \mu_J(f) = f - f_J.$$

To get (17), write

$$\varepsilon_J = f - \mu_J(f) = \gamma_J(f_J) - f_J;$$

then use (16) and the formal definition (1.65) of α and β.

Let P_A be P_i conditioned on A. Use (1.56) and (1.66) to construct processes G_n and random variables T_n and B_n on some convenient triple so that:

(18)   the joint distribution of

$$T_0, G_0(T_0-), \quad B_1, \quad T_1, G_1(T_1-), \quad B_2, \ldots, \quad T_N, G_N(T_N-), B_{N+1}$$

coincides with the joint P_A-distribution of

$$\tau_{J,0}, \quad \alpha_{J,0}, \beta_{J,1}, \tau_{J,1}, \quad \alpha_{J,1}, \beta_{J,2}, \ldots, \tau_{J,N}, \quad \alpha_{J,N}, \beta_{J,N+1};$$

(19) $T_0, G_0, B_1, T_1, G_1, B_2, \ldots, T_N, G_N, B_{N+1}$ are independent;

(20) G_n satisfies (1.56f) for $n = 0, \ldots, N$.

Let m be the joint distribution of $T = (T_0, T_1, \ldots, T_N)$, a probability on the positive orthant of Euclidean $(N + 1)$-space, equivalent to Lebesgue measure there: (MC, 5.48). For $t = (t_0, \ldots, t_N)$, let

$$H(t) = \Sigma_{n=1}^{N+1} B_n + \Sigma_{n=0}^{N} [G_n(t_n-) - t_n].$$

Let D be the set of t such that:

$$t_n > 0 \quad \text{for } n = 0, \ldots, N; \quad \text{and} \quad \Sigma_{n=0}^{N} t_n \leq 2; \quad \text{and} \quad \text{Prob} \{H(t) \leq \tfrac{1}{2}\} > 0.$$

Use Fubini on (16–19):

$$P_A\{\varepsilon_J \leq \tfrac{1}{2} \text{ and } f_J \leq 2\} = \int_{t \in D} \text{Prob}\,\{H(t) \leq \tfrac{1}{2}\}\, m(dt).$$

Now (14) implies $m(D) > 0$.

For a vector $t = (t_0, \ldots, t_N)$, let $4t = (4t_0, \ldots, 4t_N)$. Let m_4 be the distribution of $4T$. Then m_4 is equivalent to m; let ρ be a strictly positive Radon Nikodym derivative of m_4 with respect to m. This can be written down explicitly, but the formula confuses things. Use (20) to check $H(4t) \geq H(t)$. Now compute.

$$P_A\{\varepsilon_J \leq \tfrac{1}{2} \text{ and } 4(\tau_{J,0}, \ldots, \tau_{J,N}) \in D\}$$

$$= \int_{4t \in D} \text{Prob}\,\{H(t) \leq \tfrac{1}{2}\}\, m(dt)$$

$$\geq \int_{4t \in D} \text{Prob}\,\{H(4t) \leq \tfrac{1}{2}\}\, m(dt)$$

$$= \int_{t \in D} \text{Prob}\,\{H(t) \leq \tfrac{1}{2}\}\, m_4(dt)$$

$$= \int_{t \in D} \text{Prob}\,\{H(t) \leq \tfrac{1}{2}\}\, \rho(t)\, m(dt)$$

$$> 0.$$

But $4(\tau_{J,0}, \ldots, \tau_{J,N}) \in D$ entails $f_J \leq \tfrac{1}{2}$, using (16) again. Continue with the help of (15).

$$P_A\{f \leq 1\} \geq P_A\{\varepsilon_J \leq \tfrac{1}{2} \text{ and } f_J \leq \tfrac{1}{2}\}$$

$$\geq P_A\{\varepsilon_J \leq \tfrac{1}{2} \text{ and } 4(\tau_{J,0}, \ldots, \tau_{J,N}) \in D\}$$

$$> 0.$$

But $P_i(A) > 0$ by (14). This proves (11), and with it (8). ★

4. DETERMINING THE TIME SCALE

The result (21) of this section explains some of the difficulty in (8), and part of the proof. Suppose a Markov sample function is subjected to a homogeneous change of time scale $t \to t/c$. Then c can be computed with probability 1 from a piece of sample function iff the piece is not a step function. To state this precisely, let A be the set of ω in Ω_∞ such that $X(\cdot, \omega)$ is a right continuous I-valued step function† on $[0, 1)$, with an I-valued limit

†See page 165 of *MC* for the definition.

from the left at 1, and $X(1, \omega) \in I$. Let $B = \Omega_\infty \setminus A$. For $0 < c < \infty$, define the standard stochastic semigroup P^c on I by $: P^c(t) = P(ct)$. Recall that $\mathscr{F}(1)$ is the σ-field spanned by $X(s)$ for $0 \leqq s \leqq 1$. Thus $A \in \mathscr{F}(1)$. Fix $i \in I$.

(21) Theorem.   (a)   *The probabilities P_i^c retracted to $A \cap \mathscr{F}(1)$ are mutually equivalent.*

   (b)   *There is an $\mathscr{F}(1)$-measurable function f from B to $(0, \infty)$, with $P_i^c\{f = c|B\} = 1$ for all c.*

As usual, let $Q = P'(0)$, the generator of P. Of course, the generator of P^c is cQ. Let $q(j) = -Q(j, j)$. For $j \neq k$ and $q(j) < \infty$, let $\Gamma(j, k) = Q(j, k)/q(j)$. Let $\Gamma(j, k) = 0$ otherwise. Let X visit states $\xi_0, \xi_1, \ldots$ with holding times $\tau_0, \tau_1, \ldots$. The formal definition is before $(MC, 9.48)$. Of course, ξ_1 is defined only on part of the sample space, perhaps a part with probability 0.

(22)   With respect to P_i^c, the process $\xi_0, \xi_1, \ldots$ is a partially defined discrete time Markov chain, with starting state i and stationary substochastic transitions Γ; given $\xi_0 = i, \xi_1 = i_1, \ldots, \xi_n = i_n$, the random variables $\tau_0, \tau_1, \ldots, \tau_n$ are conditionally independent and exponentially distributed, with parameters $cq(i), cq(i_1), \ldots, cq(i_n)$, respectively.

For a discussion of (22), see $(MC, 9.48)$.

PROOF OF (a).   Define a random variable $N = 0, 1, \ldots$ on A, as follows: $\{N = 0\} = \{\tau_0 > 1\}$. Let $n = 1, 2, \ldots$. Then $N = n$ iff $\xi_0, \ldots, \xi_n$ are defined and

$$\tau_0 + \cdots + \tau_{n-1} = 1$$

or

$$\tau_0 + \cdots + \tau_{n-1} < 1 < \tau_0 + \cdots + \tau_n.$$

Thus, $N + 1$ is the number of steps in the initial step-function piece of X retracted to $[0, 1]$. The last step is allowed to degenerate to a point. See Figure 4 for the nondegenerate case $N = 2$. Check $\{N = n\} \in \mathscr{F}(1)$, although

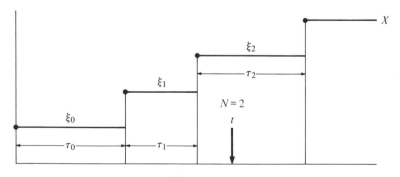

Figure 4.

$\tau_0 + \cdots + \tau_n$ is not $\mathscr{F}(1)$-measurable, even on $\{N = n\}$. Moreover, A is the union of $\{N = n\}$. On $\{N = n\}$, the σ-field $\mathscr{F}(1)$ is spanned by $\xi_0, \ldots, \xi_n$ and $\tau_0, \ldots, \tau_{n-1}$. For $i_0, i_1, \ldots, i_n$ in I, the number

$$P_i^c\{\xi_0 = i_0, \xi_1 = i_1, \ldots, \xi_n = i_n\}$$

does not depend on c, by (22). As c varies, the conditional joint P_i^c-distributions of $(\tau_0, \ldots, \tau_n)$ given

$$\{\xi_0 = i_0, \xi_1 = i_1, \ldots, \xi_n = i_n\}$$

are mutually equivalent, again by (22). Therefore, the conditional joint P_i^c-distributions of $(\tau_0, \ldots, \tau_{n-1})$ given

$$\{\xi_0 = i_0, \xi_1 = i_1, \ldots, \xi_n = i_n \text{ and } N = n\}$$

must also be mutually equivalent.   ★

PROOF OF (b) IN A SPECIAL CASE. Suppose $q(j) < \infty$ and $\Sigma_k Q(j, k) = 0$ for all j in I. Let W be the set where ξ_n is defined for all n. Then $P_i^c(W) = 1$ for all c, and $W \cap B$ is $\mathscr{F}(1)$-measurable, being the set where all ξ_n are defined and $\Sigma_{n=0}^{\infty} \tau_n \leq 1$. See Figure 5.

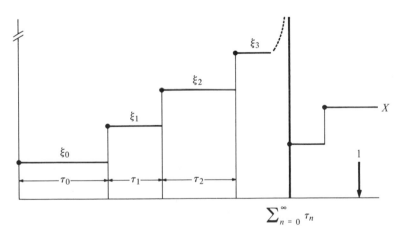

Figure 5.

Relation (22) shows that with respect to P_i^c: given $\xi_0, \xi_1, \ldots$, the random variables $q(\xi_0)\tau_0, q(\xi_1)\tau_1, \ldots$ are independent and exponentially distributed with parameter c. The same is therefore true unconditionally, and by the strong law

$$(23) \qquad \lim_{n \to \infty} \frac{1}{n} \Sigma_{m=0}^{n-1} q(\xi_m)\tau_m = \frac{1}{c}$$

with P_i^c-probability 1, for all c. Let f be 1 divided by the limit on the left side of (23), when this exists, and $f = 1$ otherwise, Let $f = 1$ off W. You should check that f retracted to B is $\mathcal{F}(1)$-measurable, although original f is not; and

(24) $P_i^c\{f = c|B\} = 1$ for all c.   ★

PROOF OF (b) IN THE GENERAL CASE‡.   Let $\tau(\omega)$ be the sup of all $t \geq 0$ such that $X(\cdot, \omega)$ is a right-continuous I-valued step function on $[0, t)$. Let D be the set of ω such that $\tau(\omega) < \infty$ and $X(\cdot, \omega)$ has an I-valued limit from the left at $\tau(\omega)$. Let $C = \{\tau < \infty\} \setminus D$. See Figure 6. Then

(25) $B = \{C \text{ and } \tau \leq 1\} \cup \{D \text{ and } \tau < 1\} \cup \{X(1) = \varphi\}$.

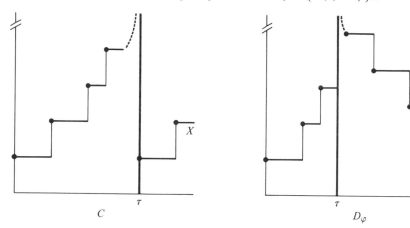

C D_φ

Figure 6.

The construction of f on $\{C \text{ and } \tau \leq 1\}$ proceeds as in the special case, and will not be considered again. The construction of f on $\{D \text{ and } \tau < 1\}$ is harder. By quasiregularity, $X(\tau+) = q\text{-lim}_{t \downarrow 0} X(\tau + t)$ exists in $\bar{I}$. For $h \in \bar{I}$, let $D_h = \{D \text{ and } X(\tau +) = h\}$. The construction is carried out separately on each piece D_h; this is legitimate because $\{D_h \text{ and } \tau < 1\} \in \mathcal{F}(1)$. Only D_φ will be considered in detail; see Figure 6. For convenience, adopt the notation of Section 1.9. Let $\mathbf{P}_i$ be the probability on the set of sample functions making the coordinate process X Markov with stationary transitions P and starting state i. Let $I = \{1, 2, \ldots\}$ and $I_n = \{1, 2, \ldots, n\}$. Let P_n be P restricted to I_n, and $q_n(j) = -P_n'(0, j, j)$. Check $(P^c)_n = (P_n)^c$, so the holding time parameter in j for P^c restricted to I_n is $cq_n(j)$.

To start the construction, let $\zeta_1 = 1$; let λ_1 be the least $t > \tau$ with $X(t) = 1$. Let ζ_2 be the least n such that X visits n on (τ, λ_1); and let λ_2 be the least

‡ The argument is only sketched. On a first reading of the book, skip to Section 5.

$t > \tau$ such that $X(t) = \zeta_2$. See Figure 7. Suppose ζ_m and λ_m defined. Then ζ_{m+1} is the least n such that X visits n on (τ, λ_m); and λ_{m+1} is the least $t > \tau$ such that $X(t) = \zeta_{m+1}$. These objects are all defined on D_φ, and

$$\tau < \cdots < \lambda_{m+1} < \lambda_m < \cdots < \lambda_1; \quad \text{while} \quad \cdots > \zeta_{m+1} > \zeta_m > \cdots > \zeta_1 = 1;$$

finally, $X(t) > \zeta_m$ for $\tau < t < \lambda_m$, while $X(\lambda_m) = \zeta_m$. For $m = 2, 3, \ldots$, let

$$(26) \qquad \theta_m = \text{Lebesgue } \{t : \lambda_m \leq t \leq \lambda_{m-1} \text{ and } X(t) = \zeta_m\}.$$

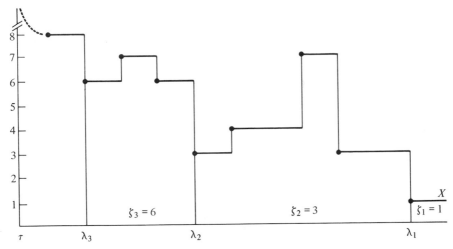

Figure 7.

Here is the main step. Let $1 < i_2 < \cdots < i_n$.

(27) With respect to $\mathbf{P}_i^c$: given D_φ and $\zeta_n = i_n, \ldots, \zeta_2 = i_2$, the random variables $\theta_n, \ldots, \theta_2$ are conditionally independent and exponentially distributed, with parameters $cq_{i_n}(i_n), \ldots, cq_{i_2}(i_2)$.

It is enough to prove (27) when $c = 1$, which eases the notation slightly. Let $t_n, \ldots, t_2$ be fixed positive numbers. Let σ be the least $t > \tau$ with $X(t) \in I_{i_2}$. Check that τ and σ are Markov times. Let F be the intersection of D_φ and the following event: on (τ, ∞), the process X reaches i_m before any $j < i_m$, for $m = n, \ldots, 2$. Check that $F \in \mathcal{F}(\sigma +)$, and $X(\sigma) = i_2$ on F. Moreover, $\theta_n, \ldots, \theta_3$ are $\mathcal{F}(\sigma +)$-measurable on F. Let G be the event: the original process X starts in i_2, and reaches 1 before any other $j < i_2$. Let θ be the time that the original process X spends in i_2 before hitting any $j < i_2$. Let $S(\omega)$ be $X[\sigma(\omega) + \cdot, \omega]$ retracted to the binary rationals. Confine ω to the set where $X[\sigma(\omega) + \cdot, \omega]$ is quasiregular. Then

$$(28) \qquad \{D_\varphi \text{ and } \zeta_n = i_n, \ldots, \zeta_2 = i_2\} = F \cap S^{-1}G$$

and

(29) $$\theta_2 = \theta(S) \quad \text{on} \quad F \cap S^{-1}G.$$

By strong Markov (9.41):

(30) $\mathbf{P_i}\{D_\varphi \text{ and } \zeta_n = i_n, \dots, \zeta_2 = i_2 \text{ and } \theta_n \geq t_n, \dots, \theta_2 \geq t_2\}$

$$= \mathbf{P_i}\{F \text{ and } \theta_n \geq t_n, \dots, \theta_3 \geq t_3 \text{ and } S \in G \text{ and } \theta \circ S \geq t_2\}$$

$$= \mathbf{P_i}\{F \text{ and } \theta_n \geq t_n, \dots, \theta_3 \geq t_3\} \cdot \mathbf{P_{i_2}}\{G \text{ and } \theta \geq t_2\};$$

(31) $\mathbf{P_i}\{D_\varphi \text{ and } \zeta_n = i_n, \dots, \zeta_2 = i_2\} = \mathbf{P_i}\{F \text{ and } S \in G\} = \mathbf{P_i}\{F\} \cdot \mathbf{P_{i_2}}\{G\}.$

Divide to get

(32) $\mathbf{P_i}\{\theta_n \geq t_n, \dots, \theta_2 \geq t_2 | D_\varphi \text{ and } \zeta_n = i_n, \dots, \zeta_2 = i_2\}$

$$= \mathbf{P_i}\{\theta_n \geq t_n, \dots, \theta_3 \geq t_3 | F\} \cdot \mathbf{P_{i_2}}\{\theta \geq t_2 | G\}.$$

To evaluate the second factor, consider the restriction of X to I_{i_2}. Then G is the event that the restricted process starts in i_2 and jumps to 1 on leaving i_2; on G, the variable θ coincides with the first holding time of the restricted process. So,

(33) $$\mathbf{P_{i_2}}\{\theta \geq t_2 | G\} = e^{-u}$$

where

$$u = q_{i_2}(i_2)t_2.$$

A similar argument, followed by induction, shows

(34) $$\mathbf{P_i}\{\theta_n \geq t_n, \dots, \theta_3 \geq t_3 | F\} = e^{-t}$$

where

$$t = q_{i_n}(i_n)t_n + \cdots + q_{i_3}(i_3)t_3.$$

This completes the argument for (27).

Introduce $T_m = q_{\zeta_m}(\zeta_m)\theta_m$ for $m = 2, 3, \dots.$ As (27) implies,

(35)   With respect to $\mathbf{P_i^c}$: given D_φ, the random variables $T_2, T_3, \dots$ are conditionally independent and exponentially distributed, with common parameter c.

As the strong law implies,

(36) $$\lim_{n \to \infty} \frac{1}{n}\Sigma_{m=2}^{n+1} T_m = \frac{1}{c}$$

with P_i^c-probability 1, for all c. Let f be 1 divided by the limit on the left side of (36), when this limit exists; and $f = 1$ elsewhere. In particular, $P_i^c\{f = c | D_\varphi\} = 1$ for all c.

Now another problem must be faced: T_m is not $\mathcal{F}(1)$-measurable, even on $\{D_\varphi$ and $\tau < 1\}$. Plainly, $\lambda_m \downarrow \tau$ on D_φ. Let M be the least m with $\lambda_m \leq 1$. Now M is still not $\mathcal{F}(1)$-measurable, even on $\{D_\varphi$ and $\tau < 1\}$. Despite these unfavorable conditions, the random variables $\zeta_M, \zeta_{M+1}, \ldots$ and $\lambda_M, \lambda_{M+1}, \ldots$ and $\theta_{M+1}, \theta_{M+2}, \ldots$ and $T_{M+1}, T_{M+2}, \ldots$ are all $\mathcal{F}(1)$-measurable on $\{D_\varphi$ and $\tau < 1\}$. To see this, begin by checking that ζ_M is the least n visited by X on $(\tau, 1]$, and λ_M is the least $t > \tau$ with $X(t) = \zeta_M$; look at Figure 7. Inductively, ζ_{M+m+1} is the least n visited by X on (τ, λ_{N+m}), and λ_{M+m+1} is the least $t > \tau$ with $X(t) = \zeta_{M+m+1}$. Consequently, these random variables are all $\mathcal{F}(1)$-measurable on $\{D_\varphi$ and $\tau < 1\}$. Further,

$$\theta_{M+m} = \text{Lebesgue}\,\{t : \lambda_{M+m} \leq t \leq \lambda_{M+m-1} \text{ and } X(t) = \zeta_{M+m}\}$$

and

$$T_{M+m} = q_{\zeta_{M+m}}(\zeta_{M+m})\theta_{N+m}$$

are $\mathcal{F}(1)$-measurable on $\{D_\varphi$ and $\tau < 1\}$, for $m = 1, 2, \ldots$. The limit on the left side of (36) is unaffected by shifting the T's, and this proves f is $\mathcal{F}(1)$-measurable on $\{D_\varphi$ and $\tau < 1\}$.

The construction on $\{D_1$ and $\tau < 1\}$ is very similar. Begin by setting $\zeta_1 = 2$, and letting λ_1 be the least $t > \tau$ with $X(t) = \zeta_1$. Define the remaining ζ's and λ's by the same induction, but with $n \geq 2$. Thus, ζ_{m+1} is the least $n \geq 2$ such that X visits n on (τ, λ_m), and λ_{m+1} is the least $t > \tau$ with $X(t) = \zeta_{m+1}$. Revise (27) thus. Let $2 < i_2 < \cdots < i_n$; let P_m be the restriction of P to $\{2, \ldots, m\}$, rather than $\{1, \ldots, m\}$; and let $q_m(j) = -P'_m(0, j, j)$. To prove (27), let F be the intersection of D_1 with the following event: on (τ, ∞), the process X reaches i_m before any j with $1 < j < i_m$, for $m = n, \ldots, 2$. Let G be the event that X starts in i_2, and reaches 2 before any j with $2 < j < i_2$. Let θ be the time X spends in i_2 before hitting any j with $2 \leq j < i_2$. To evaluate the second factor on the right in (32), restrict X to $\{2, \ldots, i_2\}$. The rest of the argument for D_1 is the same. The argument for D_j is obtained by renumbering, so as to interchange 1 and j.   ★

5. A THEOREM OF WILLIAMS

The object of this section is to prove a result (37) of Williams (1967). Adopt the notation of Section 1.9, so $I = \{1, 2, 3, \ldots\}$ and $I_n = \{1, \ldots, n\}$

and P_n is P restricted to I_n. Recall that $Q = P'(0)$ and $Q_n = P'_n(0)$. Moreover, $q_n(i) = -Q_n(i, i)$, and

$$\Gamma_n(i, j) = Q_n(i, j)/q_n(i) \quad \text{for } i \neq j$$
$$= 0 \qquad\qquad \text{for } i = j.$$

Finally,

$$\pi_n(i) = 1 - \Gamma_n(i, n)\Gamma_n(n, i).$$

(37) Theorem.   *Let $\varepsilon > 0$ and $1 \leq N < \infty$. If $Q(1, n) \geq \varepsilon$ for all $n > N$, then 1 is instantaneous and all $j > 1$ are stable.*

PROOF.   The first claim is easy (*MC*, 5.14). For the second, iterating (1.73a) shows

$$q_{N+m+1}(j) = q_N(j) \, \Pi_{n=N}^{N+m} \, \pi_{n+1}(j)^{-1}.$$

Then (1.88) shows

(38)
$$q(j) = q_N(j) \, \Pi_{n=N}^{\infty} \, \pi_{n+1}(j)^{-1}.$$

So j is stable iff

$$\Pi_{n=N}^{\infty} \, \pi_{n+1}(j)^{-1} < \infty.$$

This is equivalent to

$$\Sigma_{n=N}^{\infty} \, [1 - \pi_{n+1}(j)] < \infty.$$

Now (1.73c) implies $Q_n(1, j)$ decreases as n increases, say to $Q_\infty(1, j)$. Of course, this limit is identified in (1.90); but the identification is irrelevant here. By telescoping (1.73c),

$$\Sigma_{n=N}^{\infty} \, Q_{n+1}(1, n + 1)\Gamma_{n+1}(n + 1, j) = Q_N(1, j) - Q_\infty(1, j).$$

From the condition,

$$\Sigma_{n=N}^{\infty} \, \Gamma_{n+1}(n + 1, j) \leq \varepsilon^{-1} \, [Q_N(1, j) - Q_\infty(1, j)] < \infty.$$

But

$$1 - \pi_{n+1}(j) = \Gamma_{n+1}(j, n + 1)\Gamma_{n+1}(n + 1, j) \leq \Gamma_{n+1}(n + 1, j). \qquad \bigstar$$

6. TRANSFORMATION OF TIME

Continue in the setting of Section 1.5, but generalize the discussion as follows†. Let f be a nonnegative function on I, which does not vanish iden-

† The argument in this section is only sketched. On a first reading of the book, you should skim down to the main result (52), and then skip to Chapter 3.

tically. In Section 1.5, the function f was the indicator of a finite set. Let

$$(39) \qquad\qquad M_f(t, \omega) = \int_0^t f[X(s, \omega)] \, ds,$$

so $M_f(\infty, \omega) = \lim_{t \to \infty} M_f(t, \omega) = \infty$. Let the inf of an empty set be ∞. Let $\theta_f(\omega)$ be the inf of the t with $M_f(t, \omega) = \infty$. The function $M_f(\cdot, \omega)$ is continuous except at $\theta_f(\omega)$, where it is continuous from the left. But $M_f[\theta_f(\omega), \omega] < \infty$ is a possibility. For $t \geq M_f[\theta_f(\omega), \omega]$, let $T_f(t, \omega) = \infty$. For $t < M_f[\theta_f(\omega), \omega]$, let

$$T_f(t, \omega) \text{ be the greatest } s \text{ with } M_f(s, \omega) = t.$$

Let I_f be the set of $i \in I$ with $f(i) > 0$. Let $\partial \notin \bar{I}$ and $X(\infty) = \partial$. Let $\{\partial\}$ be an open subset of $I \cup \{\partial\}$. Let

$$(40) \qquad\qquad X_f(t, \omega) = X[T_f(t, \omega), \omega]$$

and

$$(41) \qquad\qquad Y_f(t, \omega) = X_f(t, \omega) \quad \text{when } X_f(t, \omega) \in I_f \cup \{\varphi\} \cup \{\partial\}$$
$$= \varphi \qquad\qquad \text{when } X_f(t, \omega) \in I \setminus I_f.$$

Introduce the following terminology.

(42a) *locally finitary.* Say f is *locally finitary* iff $P_i\{\theta_f = 0\} = 0$ for all $i \in I_f$. For such f, say $\omega \in \Omega_\infty$ is *exceptional* iff $\omega(0) \in I_f$ but $\theta_f(\omega) = 0$.

(42b) *finitary.* Say f is *finitary* iff $P_i\{\theta_f < \infty\} = 0$ for all $i \in I_f$. For such f, say $\omega \in \Omega_\infty$ is *exceptional* iff $\omega(0) \in I_f$ but $\theta_f(\omega) < \infty$.

Remember Ω_m from $(MC, 9.27)$. This set depends on the state space I; when this is important, I will write $\Omega_m(I)$. In a similar way, I will sometimes write $\Omega_\infty(I)$ for Ω_∞.

(43) Theorem. *Let f be locally finitary and $i \in I_f$. With respect to P_i, the process Y_f is Markov with stationary standard stochastic transitions P_f on $I_f \cup \{\partial\}$, starting from i. All sample functions are in $\Omega_m(I_f \cup \{\partial\})$. If f is finitary, or even if $M_f(\theta_f) = \infty$, then P_f is stochastic on I_f.*

PROOF.   You should argue first, as in (1.37), that

(44)   the conditional P_i-distribution of $X_f(t + \cdot)$, given $X_f(s)$ for $0 \leq s \leq t$ and $X_f(t) = j \in I_f$, coincides with the P_j-distribution of X_f.

If F is nondecreasing and continuous, let $|F|$ be the measure whose distribution function is F; let $F(A)$ be the F-image of A; and let $|A|$ be the Lebesgue measure of A. From (1.30),

$$|F(A)| = |F|(A) \quad \text{for Borel } A.$$

Of course, M_f is continuous and nondecreasing on $[0, \theta_f)$. And M_f assigns measure 0 to

$$A = \{t : 0 \leq t < \theta_f \text{ and } X(t) \notin I_f\}.$$

So

(45) $|M_f(A)| = 0.$

Let

$$B = [0, \theta_f) \setminus A$$

$$= \{t : 0 \leq t < \theta_f \text{ and } X(t) \in I_f\}.$$

Because $\omega \in \Omega_\infty$,

(46) M_f is strictly increasing on B.

So $T_f[M_f(s)] = s$ for $s \in B$. Of course, $M_f[T_f(s)] = s$ provided $T_f(s) < \theta_f$. So $T_f(t) \in B$ iff $t \in M_f(B)$. By thinking, $Y_f(t) \in I_f$ iff $X_f(t) \in I_f$ iff $T_f(t) \in B$. Consequently,

(47) $M_f(B) = \{t : X_f(t) \in I_f\} = \{t : Y_f(t) \in I_f\}.$

By (46) and (47),

(48) M_f is a strictly increasing map of B onto $\{t : Y_f(t) \in I_f\}$.

Clearly, $M_f(B) \supset [0, M_f(\theta_f)) \setminus M_f(A)$. Now (45) and (47) show $\{t : X_f(t) \in I_f\}$ is Lebesgue almost all of $[0, M_f(\theta_f))$. Consequently, $X_f(t, \omega) \in I_f \cup \{\partial\}$ for Lebesgue almost all t, for each ω. By Fubini, for Lebesgue almost all t,

(49) $P_i\{X_f(t) \in I_f \cup \{\partial\}\} = 1$ for all $i \in I_f$.

The set of t for which (49) holds has full Lebesgue measure, contains 0, and is a semigroup by (44). Then (3.15) proves that (49) holds for all t. This and (44) give the Markov property. You have to check that the sample functions are in $\Omega_m(I_f \cup \{\partial\})$. Then P_f is standard by (1.17).   ★

(50) Remark.   If f is finitary, then relative to P_f, each state in I_f is recurrent and communicates with each other state.

(51) Remark.   The argument shows $P_i\{X_f(t) = Y_f(t)\} = 1$ for $i \in I_f$ and $t \geq 0$; because (49) holds for each t.

The ambition of this section was to characterize

$$\{P_f : f \text{ is locally finitary}\}.$$

This worked out well when I is finite, so all f are finitary; but did not seem to have a neat solution in general. Consequently

$$\{P_f : f \text{ is finitary}\}$$

will be characterized. The result was suggested by (Blumenthal, Getoor and McKean, 1962) and (Knight and Orey, 1964). The characterization is in terms of the *hitting probabilities* $P(i, J)$. For the moment, drop (1.24). Let J be a finite subset of I. Let λ_J be the least t if any with $X(t) \in J$, and $\lambda_J = \infty$ if none. For $i \in I$ and finite $J \subset I \setminus \{i\}$, let $P(i, J)$ be that subprobability on J which assigns mass

$$P_i\{\lambda_J < \infty \text{ and } X(\lambda_J) = j\}$$

to $j \in J$. The main result (52) can now be stated. Assumption (1.24) is back in force. Suppose further that I has at least two elements, so no i is absorbing.

(52) Theorem. *Let H be a subset of I, having at least two elements. Let R be a standard stochastic semigroup on H. Then $R = P_f$ for some finitary f iff $R(i, J) = P(i, J)$ for all $i \in H$ and finite $J \subset H \setminus \{i\}$. In this case, $I_f = H$ and f is unique.*

The proof of (52) is deferred, for some lemmas. The following fact will not be used, but proving it may assist your understanding of (52).

(53) Fact. Let f be locally finitary. Let $i \in I$, and let J be a finite subset of $I \setminus \{i\}$. Then

$$P_f(i, J) \leq P(i, J).$$

The "only if" part of (52) is contained in

(54) Lemma. *Let f be finitary. Let $i \in I_f$, and let J be a finite subset of $I_f \setminus \{i\}$. Then*

$$P_f(i, J) = P(i, J).$$

PROOF. Let ϕ_J be the least t with $Y_f(t) \in J$. Now (43) makes Y_f Markov with transitions P_f and starting state i relative to P_i. So for $j \in J$,

$$P_f(i, J)(j) = P_i\{Y_f(\phi_J) = j\}$$
$$= P_i\{X_f(\phi_J) = j\}$$
$$= P_i\{X[T_f(\phi_J)] = j\}.$$

But $T_f(\phi_J) = \lambda_J$ on $\{\theta_f = \infty\}$, by (48).   ★

WARNING. If $\theta_f < \infty$, then $\lambda_J > \theta_f$ is possible, forcing $\phi_J = \infty$.

As usual, let $Q = P'(0)$; and let $\Gamma(i,j) = Q(i,j)/-Q(i,i)$ unless $i = j$ or $Q(i,i) = -\infty$, in which case $\Gamma(i,j) = 0$. Define Q_f and Γ_f in the analogous way for P_f.

(55) Lemma. *Suppose I is finite and f strictly positive. Then*

$$Q_f(i,i) = Q(i,i)/f(i)$$

and

$$\Gamma_f(i,j) = \Gamma(i,j).$$

PROOF. Let X wait time τ in state $X(0)$, and then jump to ξ. Similarly, let Y_f wait time τ_f in $Y_f(0) = X(0)$, then jump to ξ_f. Plainly,

$$\tau_f = f[X(0)]\tau \quad \text{and} \quad \xi_f = \xi.$$

Now use $(MC, 5.45)$ on X and $(MC, 5.48)$ on Y_f.

(56) Corollary. *Suppose I is finite and f is strictly positive. Then P_f determines f.*

PROOF. Use (55). ★

It is now possible to prove the basic case of (52). Remember that (1.24) is in force, and I has at least two elements.

(57) Lemma. *Suppose I is finite, and let R be a standard stochastic semigroup on I, with $R(i,J) = P(i,J)$ for all $i \in I$ and $J = I \setminus \{i\}$. Then $R = P_f$ for some strictly positive f, which is uniquely determined.*

PROOF. If $j \in I \setminus \{i\}$, then $(MC, 5.45)$ implies

$$\Gamma(i,j) = P(i, I \setminus \{i\})(j).$$

Similarly for R. In particular, $R'(0,i,i) < 0$ for all i by $(MC, 5.29)$. Consequently, for $i \neq j$,

$$\frac{R'(0,i,j)}{R'(0,i,i)} = \frac{P'(0,i,j)}{P'(0,i,i)}.$$

Use (55) to find the unique f with $P_f'(0,i,i) = R'(0,i,i)$ for all i. Conclude

$$P_f'(0) = R'(0).$$

Then use $(MC, 5.29)$ to see $P_f = R$. ★

(58) Lemma. *Suppose f is locally finitary but not finitary. There are $i \neq j$ in I_f with $P_f(i, \{j\}) < P(i, \{j\})$.*

PROOF.   Find $k \in I_f$ such that $P_k\{\theta_f < \infty\} > 0$, as in Figure 8. Now I_f must be infinite. Fix $\omega \in \Omega_\infty$ with $0 < \theta_f(\omega) < \infty$ and $X(0, \omega) = k$. As in Figure 8, I claim:

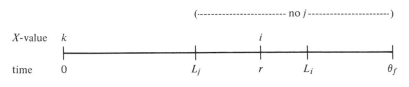

Figure 8.

(59)   There are $i \neq j$ in I_f and binary rational $r \geq 0$, all depending on ω, such that $r < \theta_f(\omega) < \infty$ and $X(r, \omega) = i$ and $X(t, \omega) \neq j$ for t with $r < t < \theta_f(\omega)$.

Indeed, let $\lambda(\omega) = q\text{-lim } X(t, \omega)$ as t increases to $\theta_f(\omega)$. For each $h \in I_f$ with $h \neq \lambda(\omega)$, let

$$L_h(\omega) = \sup \{t : 0 \leq t < \theta_f(\omega) \text{ and } X(t, \omega) = h\},$$

the sup of an empty set being $-\infty$. Of course, $L_h(\omega) < \theta_f(\omega)$. If $L_j(\omega) = -\infty$, this j and $i = k$ and $r = 0$ work. Otherwise, suppose $L_h(\omega) \geq 0$ for all h. Choose $i \neq j$ in I_f. Quasiregularity prevents $L_i(\omega) = L_j(\omega)$. Interchange i and j if necessary to get

$$0 \leq L_j(\omega) < L_i(\omega).$$

Find r with

$$L_j(\omega) < r < L_i(\omega) \quad \text{and} \quad X(r, \omega) = i,$$

as in Figure 8. Then i, j, r work. This establishes (59).

By (59) and countable additivity, there are $i \neq j$ in I_f and binary rational $r \geq 0$ such that

(60) $P_k\{r < \theta_f < \infty \text{ and } X(r) = i \text{ and } X(t) \neq j \text{ for } r < t < \theta_f\} > 0$.

Then the Markov property $(MC, 9.31)$ implies $P_i(A) > 0$, where

$$A = \{0 < \theta_f < \infty \text{ and } X(0) = i \text{ and } X(t) \neq j \text{ for } 0 < t < \theta_f\}.$$

Now Y_f does not visit j on A, so

$$P_f(i, \{j\})(j) \leq 1 - P_i(A) < 1.$$

But (1.24) forces $P(i, \{j\})(j) = 1$.   ★

For (61), let F and G be continuous and nondecreasing on $[0, \infty)$, with

$F(0) = G(0) = 0$ and $F(\infty) = G(\infty) = \infty$. Let $F^{-1}(y)$ be the greatest x with $F(x) = y$. Thus, F^{-1} is right continuous.

(61) Lemma. $(G \circ F)^{-1} = F^{-1} \circ G^{-1}$.

PROOF.   Easy.   ★

For (62) and (65), let f be finitary. Let g be a nonnegative, bounded function on I_f, which does not vanish identically. Extend g to all of I by setting $g = 0$ on $I \setminus I_f$. Let

$$L_g(t, \omega) = \int_0^t g[Y_f(s, \omega)]\, ds.$$

Delete any ω exceptional for f, as defined in (42).

(62) Lemma. $(L_g \circ M_f)(t) = \int_0^t f[X(s)]g[X(s)]\, ds.$

At ω, the left side of (62) is $L_g[M_f(t, \omega), \omega]$.

PROOF.   Since L_g and M_f are nondecreasing and absolutely continuous, so is $L_g \circ M_f$. It is enough to show that both sides of (62) have the same derivative a.e. For a discussion of absolute continuity, see Section 10.15 of MC. The derivative of the right side of (62) at almost all t is

(63) $f[X(t)]g[X(t)]$.

The chain rule works a.e. on the left side, and produces this derivative at almost all t:

(64) $g\{Y_f[M_f(t)]\}f[X(t)]$.

If $X(t) \notin I_f$, both (63) and (64) vanish. If $X(t) \in I_f$, then (46) makes $T_f[M_f(t)] = t$, so $X_f[M_f(t)] = X(t) = Y_f[M_f(t)]$, and expressions (63–64) are still equal.   ★

(65) Corollary. $(Y_f)_g = X_{fg}$, so $(P_f)_g = P_{fg}$.

PROOF.   Use (61) and (62).   ★

This generalizes (1.35). To define $(Y_f)_g$ properly, let

$$S_g(t, \omega) \text{ be the greatest } s \text{ with } L_g(s, \omega) = t.$$

Then

$$(Y_f)_g(t, \omega) = Y_f[S_g(t, \omega), \omega].$$

Because g is bounded, $L_g(\cdot, \omega)$ is finite and continuous on $[0, \infty)$. By assumption, f is finitary. And $M_f(\infty, \omega) = \infty$ because $\omega \in \Omega_\infty(I)$. So $Y_f(\cdot, \omega) \in \Omega_\infty(I_f)$, and $L_g(\infty, \omega) = \infty$. This makes S_g well-defined.

WARNING.   Here's what happens to (65) when f is locally finitary but not finitary. For simplicity, suppose f is positive everywhere, and g vanishes off the finite set G of states. Fix an ω with

$$\theta_f(\omega) < \infty \quad \text{and} \quad M_f[\theta_f(\omega), \omega] = \infty.$$

This is typical. Then

$$L_g \circ M_f(t, \omega) = M_{fg}(t, \omega) \qquad \text{for } t \leq \theta_f(\omega)$$

$$= M_{fg}[\theta_f(\omega), \omega] < \infty \quad \text{for } t \geq \theta_f(\omega).$$

There is no natural way to define $(Y_f)_g(t, \omega)$ for $t \geq M_{fg}[\theta_f(\omega), \omega]$, except by putting it in ∂. And

$$t \to (Y_f)_g(t, \omega) \quad \text{for } 0 \leq t < M_{fg}[\theta_f(\omega), \omega]$$

coincides up to a homeomorphism of time with

$$t \to X(t, \omega) \qquad \text{for } 0 \leq t < \theta_f(\omega) \text{ and } X(t, \omega) \in G.$$

By contrast,

$$t \to X_{fg}(t, \omega) \quad \text{for } 0 \leq t < \infty$$

coincides up to a homeomorphism of time with

$$t \to X(t, \omega) \qquad \text{for } 0 \leq t < \infty \text{ and } X(t, \omega) \in G.$$

Recall that $\mathscr{F}(\varepsilon)$ is the σ-field spanned by $X(s)$ for $0 \leq s \leq \varepsilon$, and remember $\mathscr{F}(0+) = \bigcap_{\varepsilon > 0} \mathscr{F}(\varepsilon)$. The next result is a minor variation on a piece of (Blumenthal, 1957).

(66) Lemma. $\mathscr{F}(0+)$ is trivial with respect to P_i.

PROOF.   By the strong Markov property $(MC, 9.41)$, for any measurable A,

$$P_i\{A|\mathscr{F}(0+)\} = P_i(A) \quad \text{with } P_i\text{-probability 1}.$$

If $A \in \mathscr{F}(0+)$,

$$P_i\{A|\mathscr{F}(0+)\} = 1_A \qquad \text{with } P_i\text{-probability 1}.$$

Choose an ω for which both equations hold:

$$P_i(A) = 1_A(\omega) = 0 \quad \text{or} \quad 1. \qquad \bigstar$$

(67) Corollary.   If f is nonnegative, $i \in I_f$, and $P_i\{\theta_f = 0\} > 0$, then $P_i\{\theta_f = 0\} = 1$.

PROOF.   Use (66). $\qquad \bigstar$

PROOF OF (52).   Let F be a finite subset of H, with at least two elements.

Then R_F and P_F are standard stochastic semigroups on F. As (54) implies, R_F and P_F have the same hitting probabilities. By (57), there is a unique positive function f_F on F, such that

$$R_F = (P_F)_{f_F}.$$

Let F and G be finite with $F \subset G \subset H$. Let ρ be the retract of f_G to F. I claim

(68) $\rho = f_F.$

Here is a preliminary assertion:

(69) $[(P_G)_{f_G}]_F = [(P_G)_F]_\rho.$

To argue (69), use (65) separately on the left side and the right side to see that each one is $(P_G)_h$, where $h = \rho = f_G$ on F and $h = 0$ on $G \setminus F$. Namely, use G for I and P_G for P. On the left side, use f_G for f; put $g = 1$ on F, and $g = 0$ on $G \setminus F$. On the right side, let $f = 1$ on F and $f = 0$ on $G \setminus F$, while $g = \rho$; as (65) requires, g gets extended to vanish on $G \setminus F$. This establishes (69). To get (68), compute:

$$(P_F)_{f_F} = R_F \qquad \text{construction}$$

$$= (R_G)_F \qquad \text{(65) or (1.35)}$$

$$= [(P_G)_{f_G}]_F \qquad \text{construction}$$

$$= [(P_G)_F]_\rho \qquad \text{(69)}$$

$$= (P_F)_\rho \qquad \text{(65) or (1.35).}$$

Now (56) forces (68).

Relation (68) shows there is a unique positive function f on H, such that f_F is the retraction of f to F. Extend f to be 0 on $I \setminus H$. The problem is to show that f is finitary and $P_f = R$. It is not clear at this point that f is even locally finitary. Fix a sequence F_n of finite subsets of H which swell to H, and fix $i \in F_1$. Abbreviate $f_n = f_{F_n}$ and $g_n = f 1_{F_n}$: so $g_n = f = f_n$ on F_n, and $g_n = 0$ off F_n. Thus, g_n is bounded and $g_n \uparrow f$. Let $Y_n = Y_{g_n}$, so Y_n has transitions

$$P_{g_n} = (P_{F_n})_{f_n} = R_{F_n}.$$

The first equality is from (65), and the second from construction. By (65),

$$Y_n = (Y_{n+1})_{F_n}.$$

On some convenient triple, construct a Markov chain Z with stationary transitions R, starting from i, having sample functions in $\Omega_\infty(H)$. Let $Z_n = Z_{F_n}$. Then $\{Y_1, Y_2, \ldots\}$ is distributed like $\{Z_1, Z_2, \ldots\}$; indeed, $(Y_{n+1})_{F_n} = Y_n$ just as $(Z_{n+1})_{F_n} = Z_n$; while Y_n and Z_n are Markov with transitions R_{F_n}, starting from i, having right continuous step functions as sample functions. As (1.48) and (1.53) imply, $Y_n(t)$ converges in P_i-probability and in q-lim with P_i-probability 1 to an H-valued random variable $\hat{Y}(t)$, which is defined a.e. This variable will be identified later. In any case, the process $\hat{Y}$ is Markov with stationary transitions R and starting state i, relative to P_i: because $R = \lim_n R_{F_n}$ by (1.53).

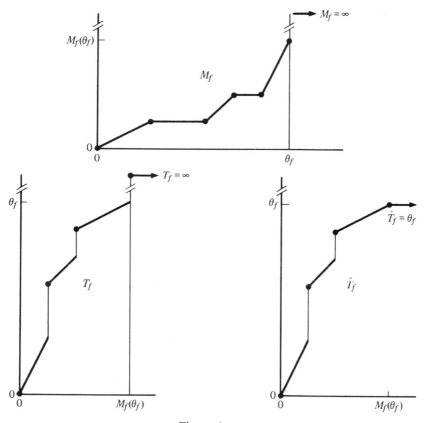

Figure 9.

Let
$$\Omega_e = \{\omega : \theta_f(\omega) < \infty \text{ and } M_f(\theta_f(\omega)) < \infty\}.$$

Let $\hat{T}_f(t, \omega) = T_f(t, \omega)$ when $\omega \notin \Omega_e$ or $\omega \in \Omega_e$ but $t < M_f(\theta_f(\omega))$). Let $\hat{T}_f(t, \omega) = \theta_f(\omega)$ when $\omega \in \Omega_e$ and $t \geq M_f(\theta_f(\omega))$). See Figure 9. By monotone convergence, $M_{g_n}(t, \omega) \uparrow M_f(t, \omega)$ for all t and ω; so $T_{g_n}(t, \omega) \downarrow \hat{T}_f(t, \omega)$. Let
$$\hat{X}(t, \omega) = X[\hat{T}_f(t, \omega), \omega].$$

Now $Y_n(t) = X[T_{g_n}(t)]$. So the quasiregularity of X implies

$$\hat{Y}(t) = \hat{X}(t) \quad \text{with } P_i\text{-probability 1.}$$

Consequently,

(70) $\hat{X}$ is Markov with stationary transitions R and starting state i, relative to P_i.

I will now argue that f is locally finitary. If not, use (67) to find $i \in H$ with $P_i\{\theta_f = 0\} = 1$. Temporarily, confine ω to $\{X(0) = i$ and $\theta_f = 0\}$. Then $M_f \equiv \infty$ on $(0, \infty)$, and $\hat{T}_f \equiv 0$ on $[0, \infty)$, and $\hat{X} \equiv i$ on $[0, \infty)$. Therefore i is absorbing for R, by (70) so for P by the condition of (52). This is false, proving f is locally finitary. I will now argue $P_i\{\Omega_e\} = 0$. Clearly,

(71) $\hat{X}(t) = X(\theta_f) \quad \text{for } t \geq M_f(\theta_f), \quad \text{on } \Omega_e.$

But $\hat{X}$ has no absorbing states by (70), so (71) makes

$$P_i\{\Omega_e \text{ and } X(\theta_f) \in I\} = 0.$$

Since $P_i\{\hat{X}(t) \in I_f\} = 1$ for all t, relation (71) makes

$$P_i\{\Omega_e \text{ and } X(\theta_f) = \varphi\} = 0.$$

The conclusion is $P_i\{\Omega_e\} = 0$. This and (51) force

(72) $P_i\{\hat{X}(t) = X_f(t) = Y_f(t)\} = 1.$

Now $R = P_f$: because Y_f has transitions R by (70, 72) and P_f by definition. And f is finitary by (58). For the uniqueness, let f_F be the retraction of f to F. Then

$$R_F = (P_f)_F = (P_F)_{f_F}$$

by (65). So f_F is pinned down by (56).   ★

Let $M(t, \omega)$ be a continuous, nondecreasing function of t, and an $\mathscr{F}(t)$-measurable function of ω. Let $T(\cdot, \omega)$ be the right continuous inverse of $M(\cdot, \omega)$. Suppose $\{X[T(t)]: 0 \leq t < \infty\}$ is a recurrent Markov chain with stationary standard stochastic transitions on a subset of I. It is then reasonable to guess that $M = M_f$ for some finitary f. This question was raised by Aryeh Dvoretzky.

As (58) shows,

$$\{P_f : f \text{ is locally finitary}\}$$

includes chains with smaller hitting probabilities. Now chains with smaller hitting probabilities can be generated by killing the original chain at finite or infinite rates on the finite states and on the infinite states. The natural theorem may then be true: $R(i, J) \leq P(i, J)$ iff R is obtained from P by transformation of time and killing.

7. THE TRANSIENT CASE

Suppose now that assumption (1.24) is withdrawn. What happens? Start with Section 1.5. To begin with, ω cannot be confined to Ω_∞, but only to Ω_m; this set was defined in (MC, 9.27). Next, $\mu_J(\infty, \omega) < \infty$ is a possibility. The proper definition for $\gamma_J(t, \omega)$ is now the inf of all s with $\mu_J(s, \omega) > t$, where the inf of an empty set is ∞. Choose $\partial \notin \bar{I}$, and let $X(\infty, \omega) = \partial$. Then X_J, defined by the same formula, is Markov with stationary standard transitions P_J on $J \cup \{\partial\}$, where ∂ is absorbing. Now P_J is substochastic on J, but stochastic on $J \cup \{\partial\}$. The rest of Chapter 1 goes through with minor and obvious changes.

Turn to Sections 2 through 6 of the present chapter. There are no significant changes in Sections 3 and 5. In Section 4, one has to redefine ζ_1 as the least n visited by X on (τ, ∞). There are mild changes to make in Section 2, as follows. If i is recurrent, restrict I to the states that can be reached from i, and proceed as before. Suppose i is transient. The total time θ spent in i is finite and exponentially distributed with parameter

$$p(i) = 1 \bigg/ \int_0^\infty P(t, i, i)\, dt,$$

relative to P_i. Theorem (1) still holds. For (2), restrict j to $Q(i, j) > 0$. There are only a finite, random number of pseudo-jumps from i to j, perhaps none. Let $\hat{\theta}(j)$ be the finite sequence of times on the $\{i\}$-scale at which these pseudo-jumps occur. Thus, $\hat{\theta}(j)$ consists of the finite $\theta_n(j)$. The problem is to describe the joint distribution of θ and the $\hat{\theta}(j)$. To this end, let $u(j)$ be the P_j-probability of hitting i. By strong Markov,

$$u(j) = \int_0^\infty P(t, j, i)\, dt \bigg/ \int_0^\infty P(t, i, i)\, dt.$$

On some convenient probability triple, construct independent Poisson processes $T(j)$ with parameter $Q(i, j)u(j)$. Construct a random variable T independent of these processes, and exponential with parameter $p(i)$. Let $\hat{T}(j)$ be the restriction of $T(j)$ to $(0, T)$. Then the joint P_i-distribution of θ and $\{\hat{\theta}(j)\}$ coincides with the joint distribution of T and $\{\hat{T}(j)\}$. The proof is about the same.

The changes for Section 6 are more serious, and will be considered in some detail†. As usual, confine ω to Ω_m rather than Ω_∞. Add the following terminology.

(73)   *adequate.* Say f is *adequate* iff for all $i \in I_f$,

$$P_i\{\theta_f < \infty \text{ and } X(t) \in I_f \text{ for some } t \geq \theta_f\} = 0.$$

For such f, say $\omega \in \Omega_m$ is *exceptional* iff $\omega(0) \in I_f$ and $\theta_f(\omega) < \infty$ but $X(t, \omega) \in I_f$ for some $t \geq \theta_f(\omega)$.

Theorem (43) survives, with the following modifications. Define M_f and θ_f as before. Let

$$\Omega_e = \{\omega : \omega \in \Omega_m \text{ and } \theta_f(\omega) < \infty \text{ and } M_f[\theta_f(\omega)] < \infty\}.$$

For $\omega \notin \Omega_e$, or $\omega \in \Omega_e$ but $t < M_f[\theta_f(\omega)]$, let

$$T_f(t, \omega) \text{ be the inf of } s \text{ with } M_f(s, \omega) > t.$$

For $\omega \in \Omega_e$ and $t \geq M_f[\theta_f(\omega)]$, let $T_f(t, \omega) = \infty$. Define X_f and Y_f by the same formulas. Then Y_f is Markov with smooth sample functions and stationary standard transitions P_f on $I_f \cup \{\partial\}$, where ∂ is absorbing. Now P_f is substochastic on I_f, but stochastic on $I_f \cup \{\partial\}$.

There is an intermediate case worth considering:

(74) $P(t, i, j) > 0$ for all $t > 0$ and i, j in I.

For (75), suppose (74) but not (1.24). That is, all i may be transient. Extend P to $I \cup \{\partial\}$ by making ∂ absorbing.

(75) Theorem.   *Suppose (74). Let $H \subset I$ have at least two elements. Let R be a standard stochastic semigroup on $H \cup \{\partial\}$, with ∂ absorbing. Then $R = P_f$ for some finitary f iff: $R(i, J) = P(i, J)$ for all $i \in H$ and finite $J \subset H \setminus \{i\}$. In this case, $I_f = H$ and f is unique.*

For the next result, do not assume (74). Extend P to $I \cup \{\partial\}$ so ∂ is absorbing.

(76) Theorem.   *Let $H \subset I$. Let R be a standard stochastic semigroup on $H \cup \{\partial\}$, with ∂ absorbing. Then $R = P_f$ for some locally finitary, adequate f iff: $R(i, J) = P(i, J)$ for all $i \in H$ and finite $J \subset H \setminus \{i\}$, and $i \in H$ is recurrent or transient for R according as i is recurrent or transient for P. Then $I_f = H$, and f is unique except at states absorbing for R.*

The recurrence condition is awkward, but the need for it will be demonstrated later, by examples (78–79). The condition can be transformed to bring out the problem more clearly. Suppose $R(i, J) = P(i, J)$ for all $i \in H$

---

† The argument is condensed. On a first reading of the book, skim to the beginning of the proof of (75), then skip to Chapter 3.

and finite $J \subset H \setminus \{i\}$. Fix $i \in H$. Suppose first $R(i, \{j\}) > 0$ for some $j \in H \setminus \{i\}$. Then i is recurrent for R iff $R(i, \{j\}) = R(j, \{i\}) = 1$, and similarly for P. But $R(i, \{j\}) = P(i, \{j\})$, and $R(j, \{i\}) = P(j, \{i\})$. So i recurrent for R iff i is recurrent for P. Next, suppose $R(i, \{j\}) = 0$ for all $j \in H \setminus \{i\}$. If i is transient for R, then i exits to ∂ for R. If i is recurrent for R, then i is absorbing for R. The recurrence condition can therefore be rephrased as follows. Suppose $i \in H$ has $R(i, \{j\}) = 0$ for all j in $H \setminus \{i\}$. Then either i exits to ∂ for R and is transient for P, or i is absorbing for R and recurrent for P. In the latter case, of course, the P-recurrent class containing i has no further state in common with H.

Theorem (52) is included in (75), which in turn is included in (76).

MODIFICATIONS FOR (75).   Lemmas (54–58) remain true if (74) holds, but (1.24) does not. The argument for (54–58) is about the same. In (58), the process Y_f does not visit j on A, but X has a positive chance of doing so; Y_f visits j off A iff X does so. For (61), drop the assumption $F(\infty) = G(\infty) = \infty$. Let $F^{-1}(y) = \infty$ for $y \geq F(\infty) = \sup F$. Then (61) still holds, by cases. You should have no trouble with (62) or (65–67).

The new difficulty occurs at the beginning of the main part of the proof: R_F and P_F may be substochastic on F. They become stochastic when extended to $F \cup \{\partial\}$. Now (54) guarantees only that R_F and P_F have the same hitting probabilities on F, not on $F \cup \{\partial\}$; so (57) does not apply immediately. Suppose F has at least two elements. Assumption (74) implies $i \in F$ is absorbing neither for R_F nor P_F. Let $F_i = F \setminus \{i\}$. Because ∂ is absorbing, I claim

(77) $P_F(i, F_i \cup \{\partial\}) = P(i, F_i)$ on F_i.

Indeed, let $j \in F_i$. Let λ be the least t if any with $X(t) \in F_i$, and $\lambda = \infty$ if none. If $\lambda < \infty$, then $\mu_F(\lambda)$ is the least t with $X_F(t) \in F_i$, and the least t with $X_F(t) \in F_i \cup \{\partial\}$. Moreover, $\gamma_F[\mu_F(\lambda)] = \lambda$. If $\lambda = \infty$, then X_F does not hit F_i, and certainly does not enter $F_i \cup \{\partial\}$ at j. But X_F has transitions P_F. So

$$P_F(i, F_i \cup \{\partial\})(j) = P_i\{\lambda < \infty \text{ and } X_F[\mu_F(\lambda)] = j\}$$
$$= P_i\{\lambda < \infty \text{ and } X[\gamma_F(\mu_F(\lambda))] = j\}$$
$$= P_i\{\lambda < \infty \text{ and } X(\lambda) = j\}$$
$$= P(i, F_i)(j).$$

This completes the argument for (77). But $P_F(i, F_i \cup \{\partial\})$ is a probability on $F_i \cup \{\partial\}$, because P_F is stochastic on $F \cup \{\partial\}$ and i is not absorbing. So,

$$P_F(i, F_i \cup \{\partial\})(\partial) = 1 - P(i, F_i)(F_i).$$

Similarly for R_F. This and (77) make

$$P_F(i, F_i \cup \{\partial\}) = R_F(i, F_i \cup \{\partial\}).$$

So (57) swings back in action, and produces f_F with $R_F = (P_F)_{f_F}$. Of course, f_F is not unique at ∂; just set it equal to 1 there. The balance of the proof does not change much.   ★

MODIFICATIONS FOR (76).   Lemma (58) remains true if (74) is not assumed, for f which are locally finitary but not adequate. Here is the proof. I will find $i \neq j$ and k in I_f and binary rational r such that $P_k(D) > 0$, where

$$D = \{r < \theta_f < \infty \text{ and } X(r) = i \text{ and } X(t) \neq j \text{ for } t \text{ with } r \leq t < \theta_f$$
$$\text{and } X(t) = j \text{ for some } t \geq \theta_f\}.$$

Indeed, X visits infinitely many states in I_f on $(\theta_f - \varepsilon, \theta_f + \varepsilon)$, for any $\varepsilon > 0$. Suppose first there is an $i \in I_f$ with $P_i(A) > 0$, where A is the event that $\theta_f < \infty$ and X visits infinitely many states in I_f on $(\theta_f, \theta_f + \varepsilon)$. Then previous reasoning works. So assume $P_i(A) = 0$ for all $i \in I_f$. Because f isn't adequate, I can find k and j^* in I_f with $P_k(B) > 0$, where

$$B = \{\theta_f < \infty \text{ and } X(t) = j^* \text{ for some } t \geq \theta_f\} \setminus A.$$

Suppose that with positive P_k-probability on B, there is an $\varepsilon = \varepsilon(\omega) > 0$ such that $X(t) \neq j^*$ on $(\theta_f - \varepsilon, \theta_f)$; then $j = j^*$ works. So assume the opposite: that θ_f is a limit from the left of $S_{j^*} = \{t : X(t) = j^*\}$ a.e. on B. Delete the exceptional ω from B. For any other j, by quasiregularity, θ_f cannot be a limit point from the left of S_j. But X visits infinitely many states of I_f on $(\theta_f - \varepsilon, \theta_f)$, provided B occurs. By countable additivity, there is a $j \in I_f \setminus \{j^*\}$ such that X visits j after visiting j^*, but before time θ_f, with positive P_k-probability on B. Thus, $P(t, j^*, j) > 0$ for $t > 0$. Using countable additivity, find $i \in I_f$ and binary rational r so that $P_k(C) > 0$, where

$$C = \{r < \theta_f < \infty \text{ and } X(r) = i \text{ and } X(t) \neq j \text{ for } r \leq t < \theta_f$$
$$\text{and } X(t) = j^* \text{ for some } t \geq \theta_f\}.$$

Use $P(t, j^*, j) > 0$ to get $P_i(D) > 0$. The balance of the argument for (58) is the same. For (54), (62), and (65) assume only that f is locally finitary and adequate. You should have no trouble extending (55–57) and (66–67).

The other new difficulty is again at the beginning of the main part of the proof. Now R_F and P_F are stochastic on $F \cup \{\partial\}$, but $P(i, F \setminus \{i\}) = 0$ guarantees only that i is absorbing for P_F or jumps directly to ∂ for P_F. The first case occurs if i is recurrent for P, the second if i is transient. Since the classification for R and P is the same, the usual argument shows $R_F = (P_F)_{f_F}$, where f_F is unique except at the absorbing states. The balance of the argument is about the same. When it comes to check that f is locally finitary, suppose by way of contradiction and (66) that

$$P_i\{\theta_f = 0\} = 1$$

for some $i \in H$. The old argument makes i absorbing for R, so recurrent for P. Let C be the P-recurrent class containing i. But i cannot lead to $H \setminus \{i\}$ for R, so $C \cap H = \{i\}$. And f vanishes off H by construction, so f is bounded on C. Now

$$P_i\{X(t) \in C \text{ for all } t\} = 1,$$

so

$$P_i\{\theta_f = \infty\} = 1.$$

This contradiction makes f locally finitary. You also have to show that

$$P_i\{\theta_f < \infty \text{ and } M_f(\theta_f) < \infty \text{ and } X(\theta_f) = j\}$$

vanishes. The old argument works for $j = \varphi$. Suppose $j \in I$ and the probability in question were positive. Because θ_f is a Markov time, Strong Markov (MC, 9.41) shows

$$P_j\{\theta_f = 0\} > 0.$$

Now an old argument makes j absorbing for R, forcing $j \in H$. This contradicts the local finitariness of f.   ★

Here are two examples on the recurrence condition in (76).

(78) Example. Let $I = \{1, 2, 3, 4\}$. Suppose the standard stochastic semigroup P on I has $Q = P'(0)$ given by

$$\begin{pmatrix} -1 & 1 & 0 & 0 \\ 1 & -1 & 0 & 0 \\ 0 & 0 & -1 & 1 \\ 0 & 0 & 1 & -1 \end{pmatrix}.$$

That is, a chain with transitions P waits in each state with holding time parameter 1, jumps from 1 to 2, from 2 to 1, from 3 to 4, and from 4 to 3. Let R be a standard stochastic semigroup on $\{1, 3, \partial\}$ for which ∂ is absorbing, while jumps from 1 to 3 and 3 to 1 are forbidden, and jumps from 1 to ∂ are encouraged. Then P and R have the same trivial hitting probabilities on $\{1, 3\}$. But $R = P_f$ for no f.   ★

(79) Example. Let $I = \{1, 2, 3, 4\}$. Suppose the standard stochastic semigroup P on I has $Q = P'(0)$ given by

$$\begin{pmatrix} -1 & 1 & 0 & 0 \\ 0 & 0 & 0 & 0 \\ 0 & 0 & -1 & 1 \\ 0 & 0 & 1 & -1 \end{pmatrix}.$$

That is, a chain with transitions $\bar{P}$ jumps from 1 to 2, 3 to 4, and 4 to 3, while 2 is absorbing. If the standard stochastic semigroup R on $\{1, 3, \partial\}$ makes each state absorbing, then R and P have the same trivial hitting probabilities on $\{1, 3\}$. But $R = P_f$ for no f. ★

3

CONSTRUCTING THE GENERAL MARKOV CHAIN

1. INTRODUCTION

In this chapter, I will construct the general Markov chain with continuous time parameter, countable state space, and stationary standard transitions. Let I be the countable state space. Let I_n be a sequence of finite subsets of I which swell to I. For each n, suppose X_n is Markov with continuous time parameter, has stationary standard transition, and has right continuous I_n-valued step functions for sample functions. Suppose that X_n is the restriction of X_{n+1} to I_n, for all n. Here is the necessary and sufficient condition for the existence of a process X whose restriction to each I_n is X_n. The sum of the lengths of the $I_2, I_3, \ldots$ -intervals occurring before X_1-time t is finite. If X is chosen with moderate care, it is automatically Markov with stationary standard transitions and smooth sample functions. Sections 1.5–6 show this construction is general. I will only do the work when I forms one recurrent class; but it is quite easy to drop this condition.

In the rest of this chapter, I will construct some examples. Section 3 exhibits a standard stochastic semigroup P with

$$P'(0, i, j) = -\infty \quad \text{for } i = j$$
$$= 0 \quad \text{for } i \neq j.$$

Section 4 recreates an example of Kolmogorov with one instantaneous state. Section 5 shows that the convergence of $P(t, i, j)$ to 0 for $i \neq j$ and $t \to 0$ is arbitrarily slow. Section 6 exposes Smith's phenomenon: a standard stochastic semigroup P with

$$P'(0, 1, 1) = -\infty$$

I want to thank Isaac Meilijson and Charles Yarbrough for their help with the post-final draft of this chapter.

and

$$\limsup_{t \to 0} P'(t, 1, 1) = \infty.$$

The state 1 is instantaneous; all other states are stable.

2.   THE CONSTRUCTION

Let $(\Omega, \mathcal{F}, \mathcal{P})$ be an abstract probability triple. Let X be a jointly measurable $\bar{I}$-valued process on $(\Omega, \mathcal{F})$. As in Section 1.5, let

$$S_J(\omega) = \{t : X(t, \omega) \in J\},$$

and suppose Lebesgue $S_J(\omega) = \infty$ for all $\omega \in \Omega$ and nonempty $J \subset I$. Let

$$\mu_J(t, \omega) = \text{Lebesgue } \{S_J(\omega) \cap [0, t]\};$$

and let $\gamma_J(\cdot, \omega)$ be the right continuous inverse of $\mu_J(\cdot, \omega)$. Let

$$X_J(t, \omega) = X[\gamma_J(t, \omega), \omega].$$

Let I be a countably infinite set. Let I_n be a sequence of finite subsets of I which swell to I, and let $i \in I_1$. For each $n = 1, 2, \ldots$, let X_n be a process on $(\Omega, \mathcal{F}, \mathcal{P})$, such that:

(1a) $X_n = (X_{n+1})_{I_n};$

(1b) X_n has right continuous step functions with values in I_n for sample functions;

(1c) Lebesgue $\{t : X_n(t, \omega) = j\} = \infty$ for all $j \in I_n$ and all $\omega \in \Omega$;

(2) X_n is Markov with stationary standard transitions, say P_n, and starting state i.

The problem is to discover when there exists an essentially I-valued Markov chain X on $(\Omega, \mathcal{F}, \mathcal{P})$ such that $X_n = X_{I_n}$ for all n. This is the converse of the problem studied in Chapter 1. Informally, the necessary and sufficient condition is: in passing from X_1 to $X_2, X_3, \ldots$ the sum of the lengths of the $I_2, I_3, \ldots$ intervals inserted to the left of any given point on the X_1-time scale must be finite. The object is to state this result precisely and prove it. For $n \leq N$, let

$$\mu_{n,N}(t, \omega) = \text{Lebesgue } \{s : 0 \leq s \leq t \text{ and } X_N(s, \omega) \in I_n\};$$

and let

$$\gamma_{n,N}(t, \omega) \text{ be the greatest } s \text{ with } \mu_{n,N}(s, \omega) = t.$$

So $\gamma_{1,n}(t)$ is the rightmost point on the n-scale corresponding to t on the 1-scale: $\gamma_{1,n}(t) - t$ is the sum of the lengths of the $I_n \setminus I_1$-intervals occurring at or before time t on the 1-scale.

(3) Lemma.   *Suppose* (1). *Let* $1 \leq m \leq n \leq N$. *Let* $0 \leq t < \infty$ *and let* $\omega \in \Omega$.

   (a) $\mu_{n,N}(t, \omega)$ *is nondecreasing with n.*

   (b) $\gamma_{m,n}(t, \omega) = \mu_{n,N}[\gamma_{m,N}(t, \omega), \omega]$.

   (c) $\gamma_{m,n}(t, \omega)$ *is nondecreasing with n.*

   (d) $\gamma_{m,N}(t, \omega) = \gamma_{n,N}[\gamma_{m,n}(t, \omega), \omega]$.

Let $\gamma_m(t, \omega) = \lim_{n \to \infty} \gamma_{m,n}(t, \omega)$, *which exists by* **(c)**.

   (e) $\gamma_m(t, \omega) = \gamma_n[\gamma_{m,n}(t, \omega), \omega]$.

   (f) $\gamma_{m,N}(t, \omega) = t + \Sigma_{n=m+1}^{N} G_{m,n}(t, \omega)$, *where* $G_{m,n}(\cdot, \omega)$ *is the right continuous distribution function of a discrete measure on* $(0, \infty)$. *This measure has infinite mass, but only finitely many jumps in* $[0, t]$ *for each t. Each jump is finite. You can define*

$$G_{m,n}(t, \omega) = \text{Lebesgue } \{s : 0 \leq s \leq \gamma_{m,n}(t, \omega) \quad and \quad X_n(s, \omega) \in I_n \setminus I_{n-1}\}.$$

PROOF.   *Claim* (a) is easy.
Claim (b).   Use (1.32c) with I_m for J and I_n for K and I_N for I.
Claim (c) follows from (a) and (b).
Claim (d).   Use (1.32d) with I_m for J and I_n for K and I_N for I.
Claim (e).   Let $N \to \infty$ in (d).
Claim (f).   Use (1.32a) with $I_n \setminus I_{n-1}$ for J and I_n for K and I_N for I; recognize $G_{m,n}$ as $\mu_{J,K} \circ \gamma_{m,n}$: so

$$G_{m,n}(t, \omega) = \text{Lebesgue } \{s : 0 \leq s \leq \gamma(t, \omega) \text{ and } X_N(s, \omega) \in I_n \setminus I_{n-1}\}$$

where

$$\gamma(t, \omega) = \gamma_{n,N}[\gamma_{m,n}(t, \omega), \omega] = \gamma_{m,N}(t, \omega) \quad \text{by (d).}$$

By definition,

$$t = \text{Lebesgue } \{s : 0 \leq s \leq \gamma_{m,N}(t, \omega) \text{ and } X_N(s, \omega) \in I_m\}.$$

Sum:

$$t + \Sigma_{n=m+1}^{N} G_{m,n}(t, \omega)$$

$$= \text{Lebesgue } \{s : 0 \leq s \leq \gamma_{m,N}(t, \omega) \text{ and } X_N(s, \omega) \in I_N\}$$

$$= \gamma_{m,N}(t, \omega). \qquad \bigstar$$

(4) Lemma.   *Suppose* (1).   *Suppose* $\gamma_1(t, \omega) < \infty$ *for all t and* ω.
   (a) $\gamma_m(t, \omega) < \infty$ *for all m, t, and* ω.

   (b) $\gamma_m(t, \omega) = t + F_m(t, \omega)$, *where* $F_m(\cdot, \omega)$ *is the right continuous distribution function of a discrete measure on* $(0, \infty)$. *This measure has infinite mass, but assigns finite mass to* $[0, t]$ *for each t. The jumps may be dense.*

(c)   *For each ω,*

$$\gamma'_m(t, \omega) = 1 \quad \text{for Lebesgue almost all } t.$$

PROOF.   *Claim* (a).   As (3a) implies,

$$\gamma_{m,n}(t, \omega) \leqq \gamma_{1,n}(t, \omega) \leqq \gamma_1(t, \omega).$$

Claim (b).   The distribution function of a sum of measures is the sum of the individual distribution functions. Use (3f) to get (b), with

$$F_m(t, \omega) = \Sigma_{n=m+1}^{\infty} G_{m,n}(t, \omega).$$

Claim (c).   Use (b) and (*MC*, 10.61).   ★

(5) Theorem.   *Suppose $\{X_n\}$ satisfies condition* (1). *Then* **(a)** *is equivalent to* **(b)**.

(a)   *There is a product measurable $\bar{I}$-valued process X on $(\Omega, \mathcal{F})$ such that*

$$X_n = X_{I_n} \quad \text{for all } n.$$

(b) $\gamma_1(t, \omega) < \infty$ *for all t and ω.*

Suppose **(b)**. *Then X can be chosen to satisfy*:

Lebesgue $\{t : X(t, \omega) = \varphi\} = 0$ *for all ω*;

$\{t : X(t, \omega) = j\}$ *is metrically perfect for all $j \in I$ and all ω*;

$q\text{-}\lim_{s \downarrow t} X(s, \omega) = X(t, \omega)$ *for all t and ω.*

This and (2) *force X to be Markov with state space I, stationary standard transitions, and smooth sample functions relative to $\mathcal{P}$: for a set of ω of inner $\mathcal{P}$-probability* 1, *the function $X(\cdot, \omega)$ is quasiregular, and the function $X(\cdot, \omega)$ retracted to the binary rationals in $[0, \infty)$ belongs to Ω_∞.*

NOTES.   (a) $\gamma_1(t) - t$ is the sum of the lengths inserted at or before time t on the 1-scale.

(b) Ω_∞ was defined after (1.24).

(c)   No null set is rejected in passing from (a) to (b).

(d) $\gamma_1(\cdot, \omega)$ is nondecreasing, so (b) has content only for large t.

(e)   I can argue that $\gamma_1(1, \omega) < \infty$ for almost all ω implies (a) after rejecting a null set. But $\mathcal{P}\{\gamma_1(1, \cdot) < \infty\} = \frac{1}{2}$ is possible.

I will leave the derivation of (5b) from (5a) to you. Here is the argument in the other direction.

(6) Lemma. *Suppose* (1). *Suppose* $\gamma_1(t, \omega) < \infty$ *for all t and ω. There is a product measurable $\bar{I}$-valued process X on $(\Omega, \mathscr{F})$ such that:*

(a) $X_n = X_{I_n}$ *for all n;*

(b) Lebesgue $\{t: X(t, \omega) = \varphi\} = 0$ *for all ω;*

(c) $\{t: X(t, \omega) = j\}$ *is metrically perfect for all $j \in I$ and all ω;*

(d) $q\text{-}\lim_{s\downarrow t} X(s, \omega)$ *for all t and ω.*

PROOF.   The proof is a long haul, which I will divide up into a number of stages.

Constructing X

Define the limiting process X as follows:

$$X(s, \omega) \in I_n \quad \text{iff} \quad s = \gamma_n(t, \omega) \text{ for some } t; \quad \text{then } X(s, \omega) = X_n(t, \omega).$$

If $s = \gamma_n(t, \omega)$ for no pair (n, t), let $X(s, \omega) = \varphi$. I have to argue that X is well defined. Suppose

$$s = \gamma_n(u, \omega) = \gamma_m(t, \omega).$$

Without loss, suppose $m \leq n$. By (3e),

$$\gamma_n(u, \omega) = \gamma_m(t, \omega) = \gamma_n[\gamma_{m,n}(t, \omega), \omega].$$

But (4b) makes $\gamma_n(\cdot, \omega)$ strictly increasing; so

$$u = \gamma_{m,n}(t, \omega);$$

(1.35) and (1a) guarantee

$$X_n(u, \omega) = X_n[\gamma_{m,n}(t, \omega), \omega] = X_m(t, \omega).$$

Product measurability and property (a)

I say

(7) $\{(s, \omega): X(s, \omega) \in I_n\}$ is product measurable.

To begin with, $\mu_{n,N}(t, \omega)$ is measurable in ω by Fubini, and continuous in t by inspection. So $\mu_{n,N}$ is product measurable. But

$$\gamma_{n,N}(t, \omega) < s \quad \text{iff} \quad \mu_{n,N}(t, \omega) > s,$$

so $\gamma_{n,N}$ is product measurable, as is

$$\gamma_n = \lim_N \gamma_{n,N}.$$

For each positive rational r, define

$$f_r(s, \omega) = \gamma_n(r, \omega) \quad \text{if } \gamma_n(r, \omega) \geq s$$

$$= \infty \qquad\qquad \text{if } \gamma_n(r, \omega) < s.$$

Clearly, f_r is product measurable; so is $f = \inf_r f_r$. But $\gamma_n(\cdot, \omega)$ is right continuous and strictly increasing, so $s = \gamma_n(t, \omega)$ for some t iff $f(s, \omega) = s$. This proves (7).

By (7) and Fubini,

$$\{s : 0 \leq s \leq t \text{ and } X(s, \omega) \in I_n\}$$

is Borel. Let

$$\mu_n(t, \omega) = \text{Lebesgue } \{s : 0 \leq s \leq t \text{ and } X(s, \omega) \in I_n\}.$$

I claim

(8) $\qquad\qquad \gamma_n(\cdot, \omega)$ is the right continuous inverse of $\mu_n(\cdot, \omega)$.

Since $\gamma_n(\cdot, \omega)$ is right continuous, it is enough to check

(9) $\qquad\qquad\qquad \mu_n[\gamma_n(t, \omega), \omega] = t.$

But

$$\{s : 0 \leq s \leq \gamma_n(t, \omega) \text{ and } X(s, \omega) \in I_n\}$$

is the $\gamma_n(\cdot, \omega)$ image of $[0, t]$. To settle (9), use (1.30) and (4b): keep in mind that $F_n(\cdot, \omega)$ is discrete. I claim

(10) $\qquad\qquad\qquad\qquad X$ is product measurable.

Indeed, (7) and Fubini make $\mu_n(t, \cdot)$ measurable. By inspection, $\mu_n(\cdot, \omega)$ is continuous. So

$$\mu_n \text{ is product measurable.}$$

Let $j \in I_n$. As (8) implies, $X(s, \omega) = j$ iff

$$X(s, \omega) \in I_n \quad \text{and} \quad X_n[\mu_n(s, \omega), \omega] = j.$$

Now use (7) and the product measurability of X_n to get (10). Clearly,

(11) $\qquad\qquad\qquad X_n(t, \omega) = X[\gamma_n(t, \omega), \omega].$

Relations (10, 8, 11) prove product measurability and (6a).

Property (b)

I claim

(12) $\qquad\qquad \gamma_n(t, \omega) - t \leq \gamma_1(t, \omega) - \gamma_{1,n}(t, \omega).$

Informally, the left side is the sum of the lengths of the $I_{n+1}, I_{n+2}, \ldots$ intervals inserted at or before time t on the n-scale; and the right side is the corresponding sum at or before time t on the 1-scale. Formally, fix n and ω; abbreviate

$$\theta(t) = \gamma_n(t, \omega) - t.$$

So θ increases by (4b). Remember $\gamma_{1,n}(t, \omega) \geq t$. By (3e),

$$\gamma_n(t, \omega) - t = \theta(t) \leq \theta[\gamma_{1,n}(t, \omega)] = \gamma_1(t, \omega) - \gamma_{1,n}(t, \omega).$$

This proves (12). Let $n \uparrow \infty$ in (12) and remember $\gamma_{1,n}(t, \omega) \uparrow \gamma_1(t, \omega)$:

(13) $\gamma_n(t, \omega) \downarrow t$ as $n \uparrow \infty$,   for all t and ω.

Use (8):

(14) $\mu_n(t, \omega) \uparrow t$ as $n \uparrow \infty$,   for all t and ω.

And

$$\text{Lebesgue } \{s : 0 \leq s \leq t \text{ and } X(s, \omega) \in I\} = \lim_n \mu_n(t, \omega) = t.$$

This settles (6b).

Property (c)

Fix s and ω, and suppose

$$X(s, \omega) = j \in I_n.$$

Then

$$s = \gamma_n(t, \omega) \quad \text{and} \quad X_n(t, \omega) = j.$$

There is an $\varepsilon > 0$ such that

$$X_n(u, \omega) = j \quad \text{for } t \leq u \leq t + \varepsilon.$$

The image under $\gamma_n(\cdot, \omega)$ of $[t, t + \varepsilon]$ is a subset of $\{u : X(u, \omega) = j\}$, which has Lebesgue measure ε by (4b) and (1.30). This image is included in the interval $[s, \gamma_n(t + \varepsilon, \omega)]$ by monotonicity. But $\gamma_n(\cdot, \omega)$ is right continuous, so the interval shrinks to s as $\varepsilon \downarrow 0$. This settles (6c).

Property (d)

Fix s and ω. Fix $j \in I_n$. Suppose $s_m \downarrow s$ with $X(s_m, \omega) = j$. Then

$$s_m = \gamma_n(t_m, \omega) \quad \text{and} \quad X_n(t_m, \omega) = j.$$

But $\gamma_n(\cdot, \omega)$ is right continuous and strictly increasing, so

$$t_m \downarrow t \quad \text{and} \quad \gamma_n(t, \omega) = s.$$

And $X_n(\,\cdot\,, \omega)$ is right continuous, so $X_n(t, \omega) = j$; that is, $X(s, \omega) = j$. Next, suppose $X(s, \omega) = j$. Then

$$s = \gamma_n(t, \omega) \quad \text{and} \quad X_n(t, \omega) = j.$$

There is an $\varepsilon > 0$ such that

$$X_n(u, \omega) = j \quad \text{for } t \leq u \leq t + \varepsilon.$$

So

$$s \leq v \leq \gamma_n(t + \varepsilon, \omega) \quad \text{and} \quad X(v, \omega) \in I_n \quad \text{imply } X(v, \omega) = j.$$

And $X(v, \omega) = j$ for v arbitrarily close to t on the right, by (c), doing (6d). ★

Miscellaneous comments

(a)   The X of (6) is unique.

(b)   Define Y as follows: fix $j \in I$; let

$$Y(t, \omega) = X(t, \omega) \quad \text{when } X(t, \omega) \in I$$

$$= j \qquad \text{when } X(t, \omega) = \varphi.$$

Then Y is jointly measurable, visits φ at almost no times, and $Y_{I_n} = X_n$. The point is, the level sets of Y aren't metrically perfect, and $Y \neq X$.

(c)   Suppose $\{X_n\}$ satisfies (1). Suppose Y is a product measurable $\bar{I}$-valued process with

$$Y_{I_n} = X \quad \text{for all } n$$

and

$$\text{Lebesgue } \{t : Y(t, \omega) = \varphi\} = 0 \quad \text{for all } \omega.$$

Define X by (6). Then

$$X(t, \omega) = j \in I \quad \text{forces} \quad Y(t, \omega) = j.$$

If (2) holds, then $\mathcal{P}\{X(t) = Y(t)\} = 1$ for each t. For $\mathcal{P}\{X(t) \in I\} = 1$.

(d)   Property (6d) follows from (6a–c).

I will argue (c), but it's a digression. Let

$$v_n(t, \omega) = \text{Lebesgue } \{s : 0 \leq s \leq t \text{ and } Y(s, \omega) \in I_n\}$$

$$\delta_n(\,\cdot\,, \omega) \text{ be the right continuous inverse of } v_n(\,\cdot\,, \omega).$$

So

$$X_n(t, \omega) = Y[\delta_n(t, \omega), \omega]$$

and

$$X_n[v_n(s, \omega), \omega] = Y(s, \omega)$$

provided $v_n(\cdot, \omega)$ increases at s: where F increases at s iff $F(t) > F(s)$ for all $t > s$.

Let

$$T_j(\omega) = \{t : Y(t, \omega) = j\}$$

and let $L_j(\omega)$ be the set of t such that for all $u > t$,

$$\text{Lebesgue } \{s : t \leq s \leq u \text{ and } Y(s, \omega) = j\} > 0.$$

I claim

$$L_j(\omega) \subset T_j(\omega).$$

Fix $t \in L_j(\omega)$. Fix n with $j \in I_n$. So $v_n(\cdot, \omega)$ increases at t. You can check

$$\text{Lebesgue } \{T_j(\omega) \setminus L_j(\omega)\} = 0.$$

So there are

$$t_m \downarrow t \quad \text{with} \quad t_m \in L_j(\omega) \cap T_j(\omega).$$

And $v_n(\cdot, \omega)$ increases at t_m. Now

$$
\begin{aligned}
Y(t, \omega) &= X_n[v_n(t, \omega), \omega] \\
&= \lim_m X_n[v_n(t_m, \omega), \omega] \quad \text{by (1b)} \\
&= \lim_m Y(t_m, \omega) \\
&= j.
\end{aligned}
$$

Define

$$
\begin{aligned}
Z(t, \omega) = j \in I \quad &\text{iff} \quad t \in L_j(\omega) \\
= \varphi \quad &\text{iff} \quad t \notin \bigcup_{j \in I} L_j(\omega).
\end{aligned}
$$

Then Z satisfies (6), so $Z = X$.   ★

Back to work

(15) Lemma. *Let G be a Borel subset of $[0, \infty)$ which contains 0 and is closed under addition. Suppose $[0, \infty) \setminus G$ has Lebesgue measure 0. Then $G = [0, \infty)$.*

PROOF. Let $t > 0$. For Lebesgue almost all s in $(0, t)$, both s and $t - s$ are in G. Find one such s. Then $t = s + (t - s) \in G$.   ★

PROOF OF (5).   Suppose (1–2) and (5b). Suppose also $I_1 = \{i\}$: this involves no loss of generality, by (3e). Construct X by (6). Let

$$P(t, i, j) = \mathscr{P}\{X(t) = j\} \quad \text{for all } j \in I.$$

It is not clear at this stage that P is a semigroup, or that $\Sigma_{j \in I} P(t, i, j) = 1$. You should check

$$\gamma_{1,n} \quad \text{has stationary independent increments:}$$

remember $I_1 = \{i\}$ and use (1.64*). Consequently,

(16) γ_1 has stationary, independent increments.

By (4c) and Fubini,

$$\lim_{t \to 0+} [\gamma_1(s + t, \omega) - \gamma_1(s, \omega)]/t = 1 \quad \text{for } \mathscr{P}\text{-almost all } \omega,$$

for Lebesgue almost all s. By the stationarity part of (16), all s are the same. So

(17) $\lim_{t \to 0} \gamma_1(t, \omega)/t = 1 \quad \text{for } \mathscr{P}\text{-almost all } \omega.$

Use (8) to get

(18) $\lim_{t \to 0} \mu_1(t, \omega)/t = 1 \quad \text{for } \mathscr{P}\text{-almost all } \omega.$

Remember

$$\mu_1(t, \omega) = \text{Lebesgue } \{s : 0 \leq s \leq t \text{ and } X(s, \omega) = i\} \leq t.$$

By (18) and dominated convergence,

$$\lim_{t \to 0} \frac{1}{t} \int_\Omega \mu_1(t) \, d\mathscr{P} = 1.$$

But Fubini makes

$$\int_\Omega \mu_1(t) \, d\mathscr{P} = \int_0^t P(s, i, i) \, ds,$$

so

(19) $\lim_{t \to 0} \frac{1}{t} \int_0^t P(s, i, i) \, ds = 1.$

As in (1.46): if $j \in I_n$,

(20) $\int_0^\delta P_n(t, i, j) \, dt \geq \int_0^\delta P(t, i, j) \, dt.$

Fix $\varepsilon > 0$. Use (19) to find $\delta > 0$ with

$$\frac{1}{\delta} \int_0^\delta P(t, i, i) \, dt \geq 1 - \varepsilon.$$

Use (20) to get

$$\frac{1}{\delta} \int_0^\delta P_n(t, i, i) \, dt \geq 1 - \varepsilon.$$

Use (1.4) on P_n, with time rescaled by δ, to get

$$P_n(t, i, i) \geq 1 - 2\varepsilon \quad \text{for } 0 \leq t \leq \delta.$$

That is,

(21) $\lim_{t \to 0} P_n(t, i, i) = 1$ uniformly in n.

The next job is to get (21) with general j in place of i. To do this, introduce sets $J_1, J_2, \ldots$ with

$$\{j\} = J_1 \subset J_2 \subset \cdots \subset I$$

and $J_m = I_m$ for $m \geq M$. Here M is the least m with $j \in I_m$: so $M \geq 2$. I will construct a system $\{Y_m\}$ which satisfies (1–2), with j for i and $\{J_m\}$ for $\{I_m\}$; the new semigroups will be different for $m < M$, but are the P_m for $m \geq M$. For $m \geq M$, let $f_m(\omega)$ be the least t with $X_m(t, \omega) = j$; and let

$$Y_m(u, \omega) = X_m[f_m(\omega) + u, \omega] \quad \text{for } 0 \leq u < \infty.$$

For $m < M$, let

$$Y_m = (Y_{m+1})_{J_m}.$$

For $1 \leq m \leq n$, let

$$v_{m,n}(t, \omega) = \text{Lebesgue } \{s : 0 \leq s \leq t \text{ and } Y_n(s, \omega) \in I_m\}$$

$\delta_{m,n}(\cdot, \omega)$ be the right continuous inverse of $v_{m,n}(\cdot, \omega)$.

So v and δ are to Y as μ and γ are to X.

I claim

(22) $\gamma_{m,n}[f_m(\omega) + t, \omega] = f_n(\omega) + \delta_{m,n}(t, \omega)$ for $m \geq M$.

Indeed, (1.25d) implies

$$\gamma_{m,n}[f_m(\omega), \omega] = f_n(\omega).$$

So

$$\mu_{m,n}[f_n(\omega), \omega] = f_m(\omega);$$

then

$$\mu_{m,n}[f_n(\omega) + s, \omega] = f_m(\omega) + v_{m,n}(s, \omega).$$

Put $s = \delta_{m,n}(t, \omega)$ and $s = \delta_{m,n}(t, \omega) + \varepsilon$:

$$\mu_{m,n}[f_n(\omega) + \delta_{m,n}(t, \omega), \omega] = f_m(\omega) + v_{m,n}[\delta_{m,n}(t, \omega), \omega]$$
$$= f_m(\omega) + t;$$

if $\varepsilon > 0$,

$$\mu_{m,n}[f_m(\omega) + \delta_{m,n}(t, \omega) + \varepsilon, \omega] = f_m(\omega) + v_{m,n}[\delta_{m,n}(t, \omega) + \varepsilon, \omega]$$
$$> f_m(\omega) + t.$$

This proves (22).

I say $\{Y_m\}$ satisfies (1b–c), with $\{J_m\}$ for $\{I_m\}$. This is clear for $m \geq M$, and follows from (1.36) when $m < M$. I claim $\{Y_m\}$ satisfies (1a), with $\{J_m\}$ for $\{I_m\}$. This is clear for $m < M$. If $m \geq M$,

$$Y_m(t, \omega) = X_m[f_m(\omega) + t, \omega]$$
$$= X_{m+1}[\gamma_{m,m+1}(f_m(\omega) + t, \omega), \omega] \qquad \text{by (1a)}$$
$$= X_{m+1}[f_{m+1}(\omega) + \delta_{m,m+1}(t, \omega), \omega] \quad \text{by (22)}$$
$$= Y_{m+1}[\delta_{m,m+1}(t, \omega), \omega].$$

That is,

$$Y_m = (Y_{m+1})_{I_m}.$$

But $I_m = J_m$. This settles (1a). I maintain: $\{Y_m\}$ satisfies (2), with j for i and $\{J_m\}$ for $\{I_m\}$; for $m \geq M$, the transitions of Y_m coincide with the transitions P_m of X_m. If $m \geq M$, use strong Markov (MC, 9.41) on the distribution of X_m. If $m < M$, use (1.37).

Overall, the system

$$\{Y_m : m = 1, 2, \ldots\}$$

satisfies (1–2), with j replacing i and J_m replacing I_m and $J_1 = \{j\}$. I say the system $\{Y_m\}$ also satisfies (5a). If $M \leq m \leq n$, then (22) makes

$$\delta_{m,n}(t, \omega) \leq \gamma_m[f_m(\omega) + t, \omega].$$

Now use (3d) on $\{Y_m\}$:

$$\delta_{1,n}(t, \omega) = \delta_{m,n}[\delta_{1,m}(t, \omega), \omega]$$
$$\leq \gamma_m[f_m(\omega) + \delta_{1,m}(t, \omega), \omega].$$

This gets (5a). Consequently, (21) on $\{Y_m\}$ implies

(23)   For each j in I, $\lim_{t \to 0} P_n(t, j, j) = 1$ uniformly in n with $j \in I_n$.

For each pair, j, k in I, estimate $(MC, 5.9)$ implies $\{P_n(\cdot, j, k)\}$ is uniformly equi-continuous for all n with $j, k \in I_n$. These transitions are all bounded by 1. By Arzela-Ascoli (Dunford and Schwartz, 1958, page 266), this set of functions is pre-compact for the topology of uniform convergence on compacts. The diagonal argument $(MC, \text{Sec. } 10.12)$ produces a subsequence $\sigma = \{m\}$ such that for each pair j, k in I,

(24) $P_m(t, j, k) \to \hat{P}(t, j, k)$ uniformly on compact t-sets, as $m \to \infty$ through the subsequence σ.

I will now argue

(25) $\hat{P}$ is a standard stochastic semigroup on I.

It is not yet clear that $\hat{P} = P$. It is clear that $\hat{P}(\cdot, j, k)$ is nonnegative, continuous, and 1 or 0 at 0, according as $j = k$ or $j \neq k$. By Fatou and (24),

(26) $\Sigma_{k \in I} \hat{P}(t, j, k) \leq 1$.

Use (6b) and Fubini to get

(27) $\Sigma_{k \in I} P(t, i, k) = 1$ for Lebesgue almost all t.

Now compute: confine m to σ in line 2.

$$\int_0^t \Sigma_{k \in I} \hat{P}(s, i, k)\, ds = \lim_n \Sigma_{k \in I_n} \int_0^t \hat{P}(s, i, k)\, ds$$

$$= \lim_n \Sigma_{k \in I_n} \lim_m \int_0^t P_m(s, i, k)\, ds \quad \text{by (24)}$$

$$\geq \lim_n \Sigma_{k \in I_n} \int_0^t P(s, i, k)\, ds \qquad\qquad \text{by (20)}$$

$$= \int_0^t \Sigma_{k \in I} P(s, i, k)\, ds$$

$$= t \qquad\qquad\qquad\qquad\qquad\qquad \text{by (27).}$$

Use (26):

(28) $\Sigma_{k \in I} \hat{P}(t, i, k) = 1$ for Lebesgue almost all t.

Let G be the set of t such that

(29) $\Sigma_{k \in I} \hat{P}(t, j, k) = 1$ for all j in I.

As for (23), interchange i in (28) with general $j \in I$, and then intersect on j:

(30)   Lebesgue $\{[0, \infty) \setminus G\} = 0$.

Plainly, $0 \in G$. I say

(31) $s \in G$ and $t \in G$ imply $s + t \in G$.

Indeed, if $m \geq n$, then

$$P_m(s + t, j, k) \geq \Sigma_{h \in I_n} P_m(s, j, h) P_m(t, h, k).$$

Send m to ∞ through σ, use (24), then send n to ∞ :

(32) $\hat{P}(s + t, j, k) \geq \Sigma_{h \in I} \hat{P}(s, j, h) \hat{P}(t, h, k).$

Sum out k to get (31). Consequently, $G = [0, \infty)$ by (15). That is,

(33) $\Sigma_{k \in I} \hat{P}(t, j, k) = 1$ for all j in I and all t.

Sum out k in (32) again, and use (33). This forces equality in (32), and completes the argument for (25).

   The next problem is showing that X is Markov with stationary transitions $\hat{P}$. Fix $j_0 = i, j_1, \ldots, j_N$ in I and

$$0 = t_0 < t_1 < \cdots < t_N < \infty.$$

Let m be so large that all j_n are in I_m. Let A_m be the event

$$X_m(t_0) = j_0, X_m(t_1) = j_1, \ldots, X_m(t_N) = j_N.$$

Let A be the event

$$X(t_0) = j_0, X(t_1) = j_1, \ldots, X(t_N) = j_N.$$

Let

$$\pi = \prod_{n=0}^{N-1} \hat{P}(t_{n+1} - t_n, j_n, j_{n+1}).$$

Use (11, 13, 6d) to get

$$A \supset \{A_m \text{ occurs for infinitely many } m \in \sigma\}.$$

From (2) and (24),

$$\mathscr{P}(A_m) \to \pi \quad \text{as} \quad m \to \infty \text{ through } \sigma.$$

By Fatou,

$$\mathscr{P}(A) \geq \pi.$$

Sum out the j_n and use (33) to see that equality holds. This proves X is Markov with stationary transitions $\hat{P}$. The smoothness of X now follows from (MC, 9.37).   ★

Constructing $\{X_n\}$

(34) Proposition. *For each n, let Q_n be a matrix on I_n, and suppose:*

(a) Q_n *is the generator of standard stochastic semigroup;*

(b) $Q_n = (Q_{n+1})_{I_n}$;

(c) *for $j \neq k$ in I_n there is a sequence $j_0 = j, j_1, \ldots, j_m = k$ in I_n, with no repeated terms, and $Q(j_0, j_1), \ldots, Q(j_{m-1}, j_m)$ all positive.*

Then on some convenient probability triple $(\Omega, \mathscr{F}, \mathscr{P})$, there is a sequence of processes X_n satisfying (1–2), with $Q_n = P'_n(0)$.

FIRST PROOF.   Use (1.85).   ★

SECOND PROOF.   For each n, use $(MC, 5.45)$ to make a process Y_n which is Markov with starting state i, has stationary transitions generated by Q_n, and has right-continuous I_n-valued step functions for sample functions. For $m \leq n$, let

$$Y_{m,n} = (Y_n)_{I_m}.$$

By (1.35),

$$Y_{m,n+1} = (Y_{n,n+1})_{I_m} \quad \text{for } 1 \leq m \leq n.$$

And the distribution of $Y_{n,n+1}$ coincides with that of Y_n, by (b) and $(MC, 5.29)$. So the joint distribution of

$$\{Y_{m,n} : 1 \leq m \leq n\}$$

coincides with the joint distribution of

$$\{Y_{m,n+1} : 1 \leq m \leq n\}.$$

Kolmogorov $(MC, 10.53)$ makes a probability triple $(\Omega, \mathscr{F}, \mathscr{P})$, and a sequence of processes X_n such that : (1b) holds, and for each n the joint distribution of

$$\{X_m : 1 \leq m \leq n\}$$

coincides with that of

$$\{Y_{m,n} : 1 \leq m \leq n\}.$$

So (2) holds. By construction,

$$Y_{n,n+1} = (Y_{n+1,n+1})_{I_n}.$$

So

$$\mathscr{P}\{X_n = (X_{n+1})_{I_n}\} = 1.$$

By deleting a $\mathscr{P}$-null set of ω, you get (1a). Use (c), and $(MC, 1.56, 5.48)$: for $\mathscr{P}$-almost all ω, the sample function $X_n(\cdot, \omega)$ visits $j \in I_n$ infinitely often; so

$$\{t : X_n(t, \omega) = j \in I_n\} \quad \text{is unbounded.}$$

Use (1.22): by deleting a $\mathscr{P}$-null set of ω, you get 1b).   ★

NOTE. Condition (a) is algebraic: see $(MC, 5.29)$. So is (b): you can compute $(Q_{n+1})_{I_n}$ by iterating (1.73). You might review (1.85–86) at this point.

Suppose $\{Q_n\}$ satisfies (34). Construct $(\Omega, \mathscr{F}, \mathscr{P})$ and $\{X_n\}$ to satisfy (1–2). When does (5b) hold, at least on $\mathscr{P}$-almost all of Ω? Here is a partial answer. Make the following definitions, for $j, k \in I_n$: remember that $i \in I_n$ is the starting state.

$$q_n(j) = -Q_n(j, j).$$

$$\Gamma_n(j, k) = Q_n(j, k)/q_n(j) \quad \text{for } j \neq k \text{ with } q_n(j) > 0$$

$$= 0 \quad \text{elsewhere.}$$

$\pi_n(j, k)$ is the probability that a discrete time chain which starts from j and moves according to Γ_n visits k before returning to j.

$$\sigma(n, j) = \frac{\pi_n(i, j)q_n(i)}{\pi_n(j, i)q_n(j)}.$$

$$\sigma(n) = \Sigma_j \{\sigma(n, j) : j \in I_n \setminus I_{n-1}\}.$$

$$\sigma = \Sigma_{n=2}^{\infty} \sigma(n).$$

(35) Proposition. *Suppose $I_1 = \{i\}$. Condition (5b) holds after deleting a $\mathscr{P}$-null set provided $\sigma < \infty$. Then*

$$E\{\gamma_1(t)\} = (1 + \sigma)t.$$

PROOF. By (3f),

$$\gamma_{1,N}(t, \omega) = t + \Sigma_{n=2}^{N} G_{1,n}(t, \omega),$$

where $G_{1,n}(t, \omega)$ is the time $X_n(\cdot, \omega)$ spends in $I_n \setminus I_{n-1}$, until $X_n(\cdot, \omega)$ has spent time t in $I_1 = \{i\}$. In the notation of (1.87), with n for N,

$$G_{1,n}(t, \omega) = \Sigma_j \{W_{i,j}(t, \omega) : j \in I_n \setminus I_{n-1}\}.$$

By (1.87d),

$$E\{G_{1,n}(t)\} = \Sigma_j \{\sigma(n, j)t : j \in I_n \setminus I_{n-1}\} = \sigma(n)t.$$

So

$$E\{\gamma_{1,N}(t)\} = t + \Sigma_{n=2}^{N} \sigma(n)t.$$

But $\gamma_{1,N}(t, \omega) \uparrow \gamma_1(t, \omega)$.   ★

NOTE. There is an example at the end of Section 1.5, with $\gamma_1(t) < \infty$ and $E\{\gamma_1(t)\} = \infty$ for all t. So the condition of (35) isn't necessary.

(36) Example. The convergence of P_n, or even of X_n, does not guarantee (5a). Let X_n move cyclically $1 \to 2 \to \cdots \to n \to 1$, waiting an independent exponential amount of time with parameter 1 in each state; except that X_1 is identically 1. You can arrange (1–2). Now X_n converges to a Poisson process; but (5b) fails. ★

(37) Example. Let X_1 be identically 1. For $n \geq 2$, let X_n start in 1; wait in 1 an exponential amount of time with parameter n; then jump to each of $2, \ldots, n$ with equal probability $1/(n-1)$; wait there an exponential amount of time with parameter 1; then return to 1. You can arrange (1–2). Of course, (5b) goes badly wrong. For $t > 0$ and $n \to \infty$:

$$P_n(t, 1, j) \to 0 \quad \text{for all } j;$$

$$P_n(t, j, j) \to e^{-t} \quad \text{for } j > 1;$$

$$P_n(t, j, k) \to 0 \quad \text{for } j > 1 \quad \text{and} \quad k \neq j.$$

This limit is non-standard. Let $Q_n = P_n'(0)$, as usual. Then

$$Q_{n+1} = \begin{pmatrix} -n & 1 & 1 & \cdots & 1 \\ 1 & -1 & 0 & \cdots & 0 \\ 1 & 0 & -1 & \cdots & 0 \\ \vdots & \vdots & \vdots & & \vdots \\ 1 & 0 & 0 & \cdots & -1 \end{pmatrix}.$$

converges to the limit Q with: $-\infty$ at the $(1, 1)$ position; 1 in all the remaining positions in row 1 and column 1; and -1 in all the remaining positions on the diagonal; and 0 elsewhere. You can argue that $Q = P'(0)$ for no standard P, which is the point. ★

The following statements have a good chance to be right. In all cases, P_n converges to a limit P and $Q_n = P_n'(0)$ converges to a limit Q. Then P may be a standard stochastic semigroup, in which case $Q = P'(0)$. If not, then P is very degenerate, and $Q = S'(0)$ for no standard S.

3. A PROCESS WITH ALL STATES INSTANTANEOUS AND NO PSEUDO-JUMPS

Let $I = \{1, 2, \ldots\}$. Let Q be the matrix on I which vanishes off the diagonal and is $-\infty$ on the diagonal. My object in this section is to prove

(38) Theorem. *There is a standard stochastic semigroup P on I with $P'(0) = Q$.*

PROOF.   Let $n = 1, 2, 3, \ldots$. Let $I_n = \{1, \ldots, n\}$. Let $\varepsilon_1 = \varepsilon_2 = 1$ and let $\varepsilon_n > 0$ with $\Sigma \, \varepsilon_n < \infty$. I will argue below that there is a sequence Q_n of matrices on I_n satisfying the conditions of (34) and:

(39)   for each pair $j \neq k$ in I there is an $n(j, k) < \infty$ such that $Q_n(j, k) = 0$ for all $n \geq n(j, k)$;

(40) $Q_n(j, j) \to -\infty$ as $n \to \infty$ for each j in I;

(41) $\dfrac{\pi_n(1, n)q_n(1)}{\pi_n(n, 1)q_n(n)} \leq \varepsilon_n$ for $n \geq 2$,   where

(42a) $q_n(i) = -Q_n(i, i)$

(42b) $\Gamma_n(i, j) = Q_n(i, j)/q_n(i)$ for $i \neq j$ and $q_n(i) > 0$
           $= 0$ elsewhere

(42c) $\pi_n(i, j)$ is the probability that a discrete time Markov chain which starts from i and moves according to Γ_n visits j before returning to i.

Condition (41) is motivated by (35). By (34), there is a probability triple $(\Omega, \mathscr{F}, \mathscr{P})$, and a sequence of Markov chains X_n on $(\Omega, \mathscr{F}, \mathscr{P})$, satisfying (1–2), starting from $i = 1$ and having transitions P_n generated by Q_n. Condition (41) guarantees (5b) after deleting a $\mathscr{P}$-null set, by (35). Then (5) produces a standard stochastic semigroup P on I, whose restriction to I_n is P_n. The following computation is authorized by (1.90).

$$P'(0) = \lim_n P'_n(0) = \lim_n Q_n = Q. \qquad\qquad \bigstar$$

Here are some preliminaries to the construction of $\{Q_n\}$. Let $n \geq 2$.

(43) Lemma.   *Let Q_n be a generator on I_n, for which I_n is a communicating class. Fix $i \in I_n$. Then there is a generator Q_{n+1} on I_{n+1}, which makes I_{n+1} a communicating class, satisfies (41–42) with $n + 1$ for n, and satisfies*

(44) $(Q_{n+1})_{I_n} = Q_n$

(45) $Q_{n+1}(i, j) = 0$ *for all $j \in I_n \setminus \{i\}$.*

PROOF.   Let $f > 0$. Informally, obtain X_{n+1} from X_n by inserting an $(n + 1)$-interval at the end of each i-interval of X_n. The inserted intervals are to have independent, exponential lengths, with common parameter f. Formally, define a matrix Q_{n+1} on I_{n+1} as follows, where j ranges over

$I_n \setminus \{i\}$ and k ranges over I_n.

$$Q_{n+1}(i, i) = Q_n(i, i)$$

$$Q_{n+1}(i, j) = 0$$

$$Q_{n+1}(i, n + 1) = -Q_n(i, i)$$

$$Q_{n+1}(j, k) = Q_n(j, k)$$

$$Q_{n+1}(j, n + 1) = 0$$

$$Q_{n+1}(n + 1, i) = 0$$

$$Q_{n+1}(n + 1, j) = f \Gamma_n(i, j)$$

$$Q_{n+1}(n + 1, n + 1) = -f.$$

The rest is algebra. You can use $(MC, 5.29)$ to check Q_{n+1} is a generator, and (1.73) to check (44). You get (41) by making f large: π_{n+1} depends only on Γ_{n+1}, which is settled. ★

(46) Lemma. *Let Q_n be a generator on I_n, for which I_n is a communicating class. Fix $i \in I_n$, and fix a number $g < Q_n(i, i)$. Then there is a generator Q_{n+1} on I_{n+1}, which makes I_{n+1} a communicating class, satisfies (41–42) with $n + 1$ for n, satisfies (44), and satisfies*

(47) $$Q_{n+1}(i, i) = g.$$

NOTE. $g < 0$.

PROOF. Let $f > 0$. Informally, obtain X_{n+1} from X_n by cutting the interiors of the i-intervals of X_n and inserting $(n + 1)$-intervals. The cuts are to be independent from interval to interval. Within an interval, they are to be Poisson with the right mean. The inserted intervals are to have independent, exponential lengths with common parameter f. Formally, define a matrix Q_{n+1} on I_{n+1} as follows, where j ranges over $I_n \setminus \{i\}$ and k ranges over I_n.

$$Q_{n+1}(i, i) = g$$

$$Q_{n+1}(i, j) = Q_n(i, j)$$

$$Q_{n+1}(i, n + 1) = Q_n(i, i) - g$$

$$Q_{n+1}(j, k) = Q_n(j, k)$$

$$Q_{n+1}(j, n + 1) = 0$$

$$Q_{n+1}(n + 1, i) = f$$

$$Q_{n+1}(n + 1, j) = 0$$

$$Q_{n+1}(n + 1, n + 1) = -f.$$

The rest is algebra. You can use $(MC, 5.29)$ to check Q_{n+1} is a generator, and (1.73) to check (44). You get (41) by making f large: because π_{n+1} depends only on Γ_{n+1}, which is settled.   ★

PROOF THAT $\{Q_n\}$ EXISTS.   Construct a sequence $D_2, D_3, \ldots$ where each term is a state or a pair of distinct states. Each state i appears at infinitely many positions, all having index at least i. Each pair (i, j) appears once, in a position having index at least $\max \{i, j\}$. Let

$$Q_1 = (0) \quad \text{and} \quad Q_2 = \begin{pmatrix} -1 & 1 \\ 1 & -1 \end{pmatrix}.$$

Suppose $Q_2, \ldots, Q_n$ constructed. If $D_n = (i, j)$, use (43) to construct a generator Q_{n+1} on I_{n+1}, which makes I_{n+1} a communicating class, satisfies $(41\text{–}42)$ with $n + 1$ for n, satisfies (44), and satisfies $Q_{n+1}(i, j) = 0$. If $D_n = i$, use (46) to construct a generator Q_{n+1} on I_{n+1}, for which I_{n+1} is a communicating class, which satisfies $(41\text{–}42)$ with $n + 1$ for n, satisfies (44), and satisfies $Q_n(i, i) \leq -n$. Properties $(39\text{–}40)$ hold, because $Q_n(j, k)$ is nonincreasing with n, by (1.73).

4. AN EXAMPLE OF KOLMOGOROV

Let $I = \{1, 2, \ldots\}$. Let $a_2, a_3, \ldots$ be positive numbers with $\Sigma\, 1/a_n < \infty$. Define a matrix Q on I as follows:

$$Q = \begin{pmatrix} -\infty & 1 & 1 & 1 & \cdots \\ a_2 & -a_2 & 0 & 0 & \cdots \\ a_3 & 0 & -a_3 & 0 & \cdots \\ a_4 & 0 & 0 & -a_4 & \cdots \\ \vdots & \vdots & \vdots & \vdots & \end{pmatrix}.$$

That is, $Q(1, 1) = -\infty$. In addition, $Q(1, n) = 1$ and $Q(n, 1) = a_n$ and $Q(n, n) = -a_n$ for $n = 2, 3, \ldots$. All other entries vanish. My object in this section is to prove the following result of Kolmogorov (1951).

(48) Theorem.   *There is a standard stochastic semigroup P on I such that $Q = P'(0)$.*

PROOF.   Let $I_n = \{1, 2, \ldots, n\}$. Let $Q_1 = (0)$. For $n = 2, 3, \ldots$, define Q_n on I_n as follows:

$$Q_n = \begin{pmatrix} 1-n & 1 & 1 & 1 & \cdots & 1 \\ a_2 & -a_2 & 0 & 0 & \cdots & 0 \\ a_3 & 0 & -a_3 & 0 & \cdots & 0 \\ a_4 & 0 & 0 & -a_4 & \cdots & 0 \\ \vdots & \vdots & \vdots & \vdots & & \vdots \\ a_n & 0 & 0 & 0 & \cdots & -a_n \end{pmatrix}.$$

You should check that $\{Q_n\}$ satisfies the conditions of (34), using $(MC, 5.29)$ for (a) and (1.73) for (b). Use (34) to make a probability triple $(\Omega, \mathscr{F}, \mathscr{P})$, and a sequence of Markov chains X_n on $(\Omega, \mathscr{F}, \mathscr{P})$ which satisfy (1–2), starting from $i = 1$ and having transitions P_n generated by Q_n.

With the notation of (35),

$$\pi_n(1, n) = \frac{1}{n-1} \quad \text{and} \quad q_n(1) = n - 1$$

$$\pi_n(n, 1) = 1 \quad \text{and} \quad q_n(n) = a_n$$

$$I_n \setminus I_{n-1} = \{n\} \quad \text{and} \quad \sigma(n) = 1/a_n.$$

The test sum in (35) is $\sigma = \Sigma_{n=2}^{\infty} 1/a_n$, which is finite by assumption. So (35) guarantees (5b), after deleting a $\mathscr{P}$-null set. Use (5) to get a standard stochastic semigroup P on I with

$$P_{I_n} = P_n \quad \text{for all } n.$$

Use (1.90) to get

$$P'(0) = \lim_n P_n'(0) = \lim_n Q_n = Q. \qquad ★$$

By contrast with (48),

$$Q = \begin{pmatrix} -\infty & 1 & 1 & 1 & \cdots \\ 0 & -a_2 & 0 & 0 & \cdots \\ 0 & 0 & -a_2 & 0 & \cdots \\ 0 & 0 & 0 & -a_4 & \cdots \\ \vdots & \vdots & \vdots & \vdots & \end{pmatrix}$$

is not a generator. This result is due to Williams (1967).

5. SLOW CONVERGENCE

Let P be a standard stochastic semigroup on $I = \{1, 2, \ldots\}$. Of course, $[1 - P(t, 1, 1)]/t$ may tend to ∞ as t tends to 0. Aryeh Dvoretzky asked whether there exists some universal positive function f on $(0, \infty)$ which tends to 0 at 0, such that $[1 - P(t, 1, 1)]/f(t)$ tends to 0 as t tends to 0 for all P. A candidate f would be $f(t) = t^{\frac{1}{2}}$, which works in the finite case. The same question can be asked a little differently. Find a universal positive function g on $(0, \infty)$, which tends to 0 at 0, such that

$$\limsup_{t \to 0} [1 - P(t, 1, 1)]/g(t) < \infty$$

for all P. Then $g^{\frac{1}{2}}$ would be an f. The first formulation is more convenient, and the object of this section is to prove no f exists. This was also observed by Reuter (1969), and the elegant part of the proof below is due to him.

(49) Theorem. *Let f be a positive function on $(0, \infty)$, with $\lim_{t \to 0} f(t) = 0$. There is a standard stochastic semigroup P on I, and a sequence $t_j \downarrow 0$, with*

$$P(t_j, 1, j) \geq f(t_j) \quad \text{for } j = 2, 3, \ldots.$$

Of course, P and the sequence t_j depend on f.

PROOF. The generator Q of P will have the form

$$Q = \begin{pmatrix} -\infty & b_2 & b_3 & b_4 & \cdots \\ a_2 & -a_2 & 0 & 0 & \cdots \\ a_3 & 0 & -a_3 & 0 & \cdots \\ a_4 & 0 & 0 & -a_4 & \cdots \\ \vdots & \vdots & \vdots & \vdots & \end{pmatrix}.$$

The b's and a's are positive constants; I will choose them later. Abbreviate $r_j = b_j/a_j$ and $s_j = b_2 + \cdots + b_j$. I will impose the two constraints

(50) $$\sum_{j=2}^{\infty} r_j \leq \tfrac{1}{6}$$

and

(51) $$b_j \geq 1 \quad \text{for } j \geq 2.$$

Let $I_n = \{1, \ldots, n\}$. Let $Q_1 = (0)$. For $n \geq 2$, define the matrix Q_n on I_n as follows:

$$
Q_n = \begin{pmatrix}
-s_n & b_2 & b_3 & b_4 & \cdots & b_n \\
a_2 & -a_2 & 0 & 0 & \cdots & 0 \\
a_3 & 0 & -a_3 & 0 & \cdots & 0 \\
a_4 & 0 & 0 & -a_4 & \cdots & 0 \\
\vdots & \vdots & \vdots & \vdots & & \vdots \\
a_n & 0 & 0 & 0 & \cdots & -a_n
\end{pmatrix}.
$$

You should check that $\{Q_n\}$ satisfies the conditions of (34), using $(MC, 5.29)$ for (a) and (1.73) for (b). Use (34) to make a probability triple $(\Omega, \mathscr{F}, \mathscr{P})$, and a sequence of Markov chains X_n on $(\Omega, \mathscr{F}, \mathscr{P})$, which satisfy (1–2), with starting state $i = 1$ and transitions P_n generated by Q_n. With the notation of (35),

$\pi_n(1, n) = b_n/s_n$ and $q_n(1) = s_n$

$\pi_n(n, 1) = 1$ and $q_n(n) = a_n$

$I_n \setminus I_{n-1} = \{n\}$ and $\sigma(n) = b_n/a_n$.

The test sum in (35) is $\sigma = \Sigma_{n=2}^{\infty} b_n/a_n$, which is finite by assumption (50). So (35) guarantees (5b), after deleting a $\mathscr{P}$-null set. Use (5) to get a standard stochastic semigroup P on I, with

$$P_n = P_{I_n} \quad \text{for all } n.$$

Condition (51) makes $s_n \to \infty$, so (1.90) guarantees

$$P'(0) = \lim_n P_n'(0) = \lim_n Q_n = Q.$$

I claim

(52) $P(t, 1, 1) \geq \frac{2}{3}$ for all $t \geq 0$.

Indeed, (35) and (50) make

$$\int \gamma_1(t) \, d\mathscr{P} \leq \tfrac{7}{6} t.$$

So (1.25g) makes

$$\int [t - \mu_1(t)] \, d\mathscr{P} \leq \tfrac{1}{6} t.$$

By Fubini,

$$\int [t - \mu_1(t)] \, d\mathscr{P} = \int_0^t [1 - P(s, 1, 1)] \, ds.$$

Now use (1.4) with time rescaled by t to get (52).

For a moment, let P_n be any standard stochastic semigroup on I_n, with generator Q_n. Let $q_n(j) = -Q_n(j,j)$. Let $1 < j \leq n$. I say

(53) $$P_n(t, 1, j) = \int_0^t \Sigma_{k \neq j} P_n(t - u, 1, k) Q_n(k, j) \, e^{-q_n(j)u} \, du.$$

For example, look at the last jump to j in X_n on $[0, t]$. Or do it analytically: $P'_n(t) = P_n(t)Q_n$ by (MC, 5.25), so

$$\Sigma_{k \neq j} P_n(s, 1, k) Q_n(k, j) = P'_n(s, 1, j) + q_n(j) P_n(s, 1, j).$$

Multiply across by $\exp[-q_n(j)u]$ and integrate by parts, as in (MC, 7.62). For a general discussion, see (MC, Secs. 5.3 and 7.6).

Specialize (53) to the present case:

$$P_n(t, 1, j) = b_j \int_0^t P_n(t - u, 1, 1) e^{-a_j u} \, du.$$

Let $n \to \infty$ and use (1.53a):

(54) $$P(t, 1, j) = b_j \int_0^t P(t - u, 1, 1) e^{-a_j u} \, du.$$

So far, there is no control on the speed with which $P(t, 1, 1)$ tends to 1 as t tends to 0, except that the derivative is infinite. This control is obtained by proper choice of the parameters. Put $t_1 = 1$. For $j = 2, 3, \ldots$, suppose:

(55) $$0 < t_j \leq 3^{-j} \quad \text{and} \quad t_j < t_{j-1};$$

(56) $$f(t_j) \leq 2 \cdot 3^{-j-1} (1 - e^{-1});$$

(57) $$a_j = 1/t_j;$$

(58) $$r_j = b_j/a_j = 3^{-j}.$$

Then (50–51) are satisfied. Check this calculation.

$$P(t_j, 1, j) = b_j \int_0^{t_j} P(t_j - u, 1, 1) e^{-a_j u} \, du \qquad (54)$$

$$\geq \tfrac{2}{3} b_j \int_0^{t_j} e^{-a_j u} \, du \qquad (52)$$

$$= \tfrac{2}{3} r_j (1 - e^{-a_j t_j}) \qquad \text{calculus}$$

$$= \tfrac{2}{3} r_j (1 - e^{-1}) \qquad (57)$$

$$= 2 \cdot 3^{-j-1} (1 - e^{-1}) \qquad (58)$$

$$\geq f(t_j) \qquad (56). \qquad \bigstar$$

(59) Remark. I can argue from (2.2) that $\Sigma\, b_n/a_n = \infty$ prevents Q from being a generator. According to Reuter (1969), the matrix Q generates exactly one standard stochastic semigroup, but a continuum of standard substochastic semigroups. By suitable choice of the parameters, Reuter is able to make $t^{-1}[1 - P(t, 1, 1)]$ tend to ∞ arbitrarily slowly as $t \to 0$, while P stays stochastic. Namely, let g be a positive function on $(0, \infty)$, with $\lim_{t \to 0} g(t) = \infty$. For suitable choice of $\{a_j\}$ and $\{b_j\}$,

$$1 - P(t, 1, 1) \leq tg(t) \quad \text{for all small } t,$$

and P is a standard stochastic semigroup.

(60) Remark. According to Jurkat (1960), for any standard stochastic semigroup P,

$$\lim_{t \to 0} tP'(t, 1, 1) = 0.$$

This makes an interesting contrast to (49), and (61) below.

QUESTION. Are there universal K and ε such that

$$f(t) - 1 \leq tf'(t) \leq K[1 - f(t)]^2 \quad \text{if } f(t) \geq 1 - \varepsilon?$$

6. SMITH'S PHENOMENON

Let P be a standard stochastic semigroup on the countable set $I = \{i, j, \ldots,\}$. As Ornstein (1964) proved, $P'(t, i, j)$ exists and is finite and continuous on $(0, \infty)$. If $i \neq j$ or $i = j$ and $P'(0, i, i) > -\infty$, then $P'(t, i, j)$ is even continuous at $t = 0$. This left open the question of the continuity of $P'(t, i, i)$ at $t = 0$ when $P'(0, i, i) = \infty$. Smith (1964) settled this question by constructing a standard stochastic semigroup P with $1 \in I$ and

(61) $P'(0, 1, 1) = -\infty$ and $\lim \sup_{t \to 0} P'(t, 1, 1) = \infty.$

My object in this section is to sketch an alternative construction for (61). I will not really use the existence of P', which is hard. Instead, I will produce a state $1 \in I$ together with a sequence $\{c_n, d_n\}$ satisfying

(62) $$0 < c_n < d_n < \tfrac{1}{2} c_{n-1},$$

and

(63) $$\frac{P(d_n, 1, 1) - P(c_n, 1, 1)}{d_n - c_n} \geq n.$$

The existence of P' and the mean value theorem then give (61). In the present construction, c_n must tend to 0 rapidly. This leads to the conjecture that $\lim P'(t, i, i) = -\infty$ for most approaches of t to 0. One formalization would be that $P'(t, i, i)$ is approximately continuous at $t = 0$. Another interesting possibility is that the present class of examples exhibits the worst oscillations of $P'(t, 1, 1)$ near 0, in some suitable sense.

Here is an outline of the construction. Begin by reviewing the process Z, discussed in $(MC$, Sec. 8.3) and sketched in Figure 1. This process has two states, 1 and 0; its distribution has two parameters, b and c. The process Z starts in 1, holds there for a time, jumps to 0, holds there for a time, then jumps to 1 and starts afresh. Normalize the sample functions to be right continuous. The holding times in 0 are all equal to the constant c. The holding times in 1 are independent and exponential, with parameter b.

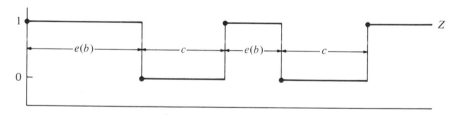

Figure 1.

Let

(64) $$f(t) = f(b, c:t) = \text{Prob}\{Z(t) = 1\}.$$

Clearly,

(65) $$f(t) = e^{-bt} \quad \text{for } t \leq c.$$

By conditioning on the first holding time,

(66) $$f(t) = e^{-bt} + \int_0^{t-c} be^{-b(t-c-s)} f(s)\, ds \quad \text{for } t \geq c.$$

This was argued in detail as $(MC, 8.39)$. In particular, the right derivative $f^+(c)$ of f at c is given by

(67) $$f^+(c) = b(1 - e^{-bc}).$$

Pause for analysis:

(68) $$1 - e^{-x} \geq x - \tfrac{1}{2}x^2 \geq \tfrac{1}{2}x \quad \text{when } 0 \leq x \leq 1.$$

So

(69) $$f^+(c) \geq \tfrac{1}{2}b^2c \quad \text{if } bc \leq 1.$$

If bc is small, then $\min f$ is close to 1, as I will argue in (81). If b^2c is large, then $f^+(c)$ is large. You can find $d > c$ but close to c, such that $f(c)$ and $f(d)$ are both close to 1; but the difference quotient of f on (c, d), namely $f(d) - f(c)$ divided by $d - c$, is large. This is the heart of my construction.

To proceed, you have to reproduce this phenomenon arbitrarily close to 0, with only one process. No Z works, in view of (65). Choose a sequence $b_n > 0$, and a sequence $\{c_n, d_n\}$ satisfying (62), such that $b_n c_n \to 0$ at a rate to be specified later, and $b_n^2 c_n \to \infty$ at a rate to be specified later. Let $d_n > c_n$ be close to c_n. Generate a sequence $Z_1 \equiv 1, Z_2, \ldots,$ of stochastic processes as follows. Let

$$0 < T_{n,1} < T_{n,2} < \cdots < T_{n,m} < \cdots$$

be a Poisson process of points, with parameter b_n, independent in n. Cut the Z_1-sample function at times $\{T_{n,m} : m = 1, 2, \ldots\}$ and insert n-intervals of length c_n for $n = 2, \ldots, N$. This produces Z_N, as in Figure 2. Now Z_N converges in a reasonable way, as $N \to \infty$, to a limiting process which exhibits

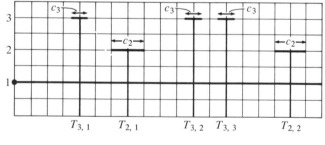

inserts for Z_3

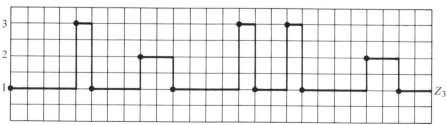

Figure 2.

Smith's phenomenon. However, this process is non-Markovian, because Z_N is non-Markovian. Modify Z_N as follows. Let f_n increase to ∞ quickly with n. Introduce new states $[n, 1], \ldots, [n, f_n]$. Let $\tau(n, j, k)$ be exponential with parameter f_n/c_n, and let these random variables be independent as n, j, k vary, and let them be independent of $\{T_{n,m}\}$. Introduce a new process X_N, as in Figure 3, with states 1 and $[n, j]$ for $j = 1, \ldots, f_n$ and $n = 2, \ldots, N$. Let X_N move through its states in the same order as Z_N, except that a visit to n is replaced by the sequence $[n, 1], \ldots, [n, f_n]$. The kth holding time of X_N in 1 is the same as the kth holding time of Z_N in 1. The kth holding time of X_N in $[n, j]$ is $\tau(n, j, k)$. In the figure, $[n, j] = n + (j/f_n)$. Now X_N is Markov, and converges to a limiting Markov process X with stationary standard transitions P, satisfying (63).

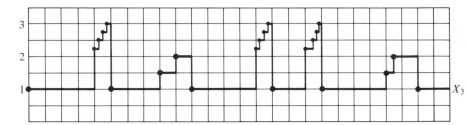

Figure 3.

It is technically easier to argue in a different order. Let $I_1 = (1)$, and let I_N consist of 1 and $[n, j]$ for $j = 1, \ldots, f_n$ and $n = 2, \ldots, N$. Write down a matrix Q_N on I_N:

$$Q_N(1, 1) = -(b_2 + \cdots + b_N)$$
$$Q_N(1, [n, 1]) = b_n$$
$$Q_N([n, j], [n, j]) = -f_n/c_n$$
$$Q_N([n, j], [n, j + 1]) = f_n/c_n \quad \text{for } j = 1, \ldots, f_n - 1$$
$$Q_N([n, f_n], 1) = f_n/c_n$$

all other entries in Q_N vanish.

NOTE.   The c_n appearing in (62) and in Q_N really are the same. The parameters $f_2, \ldots, f_N$ and $b_2, \ldots, b_N$ and $c_2, \ldots, c_N$ will be chosen later.

Check that $\{Q_N\}$ satisfies the conditions of (34): use (MC, 5.29) for (a) and (1.73) for (b). Use (34) to make a probability triple $(\Omega, \mathscr{F}, \mathscr{P})$, and a sequence X_N of Markov chains on $(\Omega, \mathscr{F}, \mathscr{P})$ satisfying (1–2), starting from $i = 1$ and having transitions P_N generated by Q_N. I will make

(70) $\Sigma\, b_n c_n < \infty.$

With the notation of (35),

$$\pi_N(1, [N, j]) = \frac{b_N}{b_2 + \cdots + b_N} \quad \text{and} \quad q_N(1) = b_2 + \cdots + b_N$$

$$\pi_N([N, j], 1) = 1 \quad \text{and} \quad q_N([N, j]) = f_N / c_N$$

$$\sigma(N, [N, j]) = b_N c_N / f_N$$

$$I_N \setminus I_{N-1} = \{[N, j] : j = 1, \ldots, f_N\}$$

$$\sigma(N) = b_N c_N.$$

The test sum in (35) is $\sigma = \Sigma_{n=2}^{\infty} b_n c_n$, which is finite by (70). So (5b) holds after deleting a $\mathscr{P}$-null set. And (5) creates a standard stochastic semigroup P on $I = \bigcup_N I_N$, such that

$$P_N = P_{I_N} \quad \text{for all } N.$$

The idea is to induct the parameters f, b, c and the auxiliary sequence d, so

(71) $0 < c_n < d_n < \tfrac{1}{2} c_{n-1} \quad \text{for } n = 2, \ldots, M$

and

(72)   the difference quotient of $P_M(\cdot, 1, 1)$ on (c_n, d_n) is more than n, for $n = 2, \ldots, M$.

Then (63) follows from (1.53a).

The argument is from continuity. To begin with, if $b_N c_N$ is small, then X_N looks like X_{N-1}.

(73) Lemma.   *Fix N, and $b_2, \ldots, b_{N-1}$ and $c_2, \ldots, c_{N-1}$, and $f_2, \ldots, f_{N-1}$. Then*

$$\lim_{b_N c_N \to 0} P_N(t, 1, 1) = P_{N-1}(t, 1, 1),$$

uniformly in b_N, c_N, f_N and in bounded t.

PROOF.   Here is a computation. The first line is (3f). The second line depends on the inequality

$$\gamma_{N-1, N}(t, \omega) \leq \gamma_{1, N}(t, \omega),$$

which holds by (3a). The third line uses the notation of (1.87).

$$\gamma_{N-1,N}(t) = t + \text{Lebesgue } \{s:0 \leqq s \leqq \gamma_{N-1,N}(t) \text{ and } X_N(s) = [N,j] \text{ for some } j\}$$

$$\leqq t + \text{Lebesgue } \{s:0 \leqq s \leqq \gamma_{1,N}(t) \text{ and } X_N(s) = [N,j] \text{ for some } j\}$$

$$= t + \Sigma_{j=1}^{f_N} W_{1,[N,1]}(t).$$

Use (1.87):

$$E\{\gamma_{N-1,N}(t)\} \leqq (1 + b_N c_N)t.$$

Now use (1.54), with I_{N-1} for J and I_N for I: so I depends on f_N. The uniformity in t is part of (1.54). The uniformity in b_N, c_N, f_N is free: if the convergence were not uniform, pick an exceptional parameter sequence and then use (1.54) on it.   ★

The next step is to examine X_N with time rescaled by c_N. For small c_N, but $b_N c_N$ fixed at λ_N, the process

$$\{X_N(tc_N):0 \leqq t < \infty\}$$

is nearly Y_N, where Y_N is a Markov chain with states 1 and $[N,j]$ for $j = 1, \ldots, f_N$, transitions

$$1 \rightarrow [N, 1] \rightarrow [N, 2] \rightarrow \cdots \rightarrow [N, f_N] \rightarrow 1,$$

holding time parameter λ_N in 1, and holding time parameter f_N in $[N,j]$. Let

(74) $$F_N(\lambda_N, f_N:t) = \text{Prob } \{Y_N(t) = 1\}.$$

More formally, let Λ_N be this generator on I_N:

$$\Lambda_N(1, 1) = -\lambda_N$$
$$\Lambda_N(1, [N, 1]) = \lambda_N$$
$$\Lambda_N([N,j], [N,j]) = -f_N \quad \text{for } j = 1, \ldots, f_N$$
$$\Lambda_N([N,j], [N, j+1]) = f_N \quad \text{for } j = 1, \ldots, f_N - 1$$
$$\Lambda_N([N, f_N], 1) = f_N$$

all other entries in Λ_N vanish.

Now $F_N(\lambda_N, f_N:\cdot)$ is the $1 - 1$ transition probability in the semigroup generated by Λ_N.

(75) Lemma. *Fix N and $b_2, \ldots, b_{N-1}$ and $c_2, \ldots, c_{N-1}$ and $f_2, \ldots, f_{N-1}$ and f_N. Fix $b_N c_N = \lambda_N > 0$. Then*

$$\lim_{c_N \to 0} P_N(tc_N, 1, 1) = F_N(\lambda_N, f_N:t)$$

uniformly in bounded t.

PROOF.   Consider the process $\hat{X}_N$, where $\hat{X}_N(t) = X_N(tc_N)$. Then $\hat{X}_N$ is Markov with transitions generated by $c_N Q_N$, which converges to Λ_N as $c_N \to 0$. Now use $(MC, 5.29)$ and the continuity of $Q \to e^Q$.   ★

Review definitions (64) and (74).

(76) Lemma.   *Fix N and λ_N. Then*

$$\lim_{f_N \to \infty} F_N(\lambda_N, f_N : t) = f(\lambda_N, 1 : t)$$

uniformly in bounded t.

PROOF.   This is $(MC, 8.43)$.   ★

CHOOSING THE PARAMETERS.   Let $c_1 = 1$. There is no b_1, d_1 or f_1. Let $N \geq 2$. Suppose $b_2, \ldots, b_{N-1}$ and $c_2, \ldots, c_{N-1}$ and $d_2, \ldots, d_{N-1}$ and $f_2, \ldots, f_{N-1}$ all chosen. Suppose (71–72) hold for $M = N - 1$. I will now choose b_N, c_N, d_N and f_N so (71–72) hold for $M = N$. Abbreviate

$$G_n(t) = P_n(t, 1, 1).$$

To begin with, I choose a positive number $\lambda_N < 1/N^2$ so small that: for any choice of b_N, c_N and f_N satisfying $b_N c_N = \lambda_N$,

(77)   the difference quotient of G_N on (c_n, d_n) exceeds n, for $n = 2, \ldots, N - 1$.

This uses (72) with $M = N - 1$, and (73) with t one of c_n or d_n for $n = 2, \ldots, N - 1$. Next, I use (69) with $b = \lambda_N$ and $c = 1$ to choose a small positive ε for which

(78)   the difference quotient of $f(\lambda_N, 1 : \cdot)$ on $(1, 1 + \varepsilon)$ exceeds $\lambda_N^2/3$.

Then, I use (76) to choose f_N so large that

(79)   the difference quotient of $F_N(\lambda_N, f_N : \cdot)$ on $(1, 1 + \varepsilon)$ is at least $\lambda_N^2/3$.

Now λ_N and f_N are fixed, as is ε. Let $d_N = (1 + \varepsilon)c_N$ and $b_N = \lambda_N/c_N$. So (70) holds. I only have c_N left. I choose it positive but so small that

(80a) $$d_N = (1 + \varepsilon)c_N < \tfrac{1}{2}c_{N-1}$$

and

(80b) $$\lambda_N^2/(6c_N) > N$$

and

(80c) $$G_N(d_N) - G_N(c_N) \geq \tfrac{1}{2}[F_N(\lambda_N, f_N : 1 + \varepsilon) - F_N(\lambda_N, f_N : 1)];$$

the last maneuver is legitimate by (75), with $t = 1 + \varepsilon$ or 1. Now

$$\frac{G_N(d_N) - G_N(c_N)}{d_N - c_N} = \frac{1}{c_N} \frac{G_N(d_N) - G_N(c_N)}{\varepsilon}$$

$$\geq \frac{1}{2c_N} \frac{F_N(\lambda_N, f_N : 1 + \varepsilon) - F_N(\lambda_N, f_N)}{\varepsilon} \qquad \text{by (80c)}$$

$$\geq \frac{1}{2c_N} \frac{\lambda_N^2}{3} \qquad \text{by (79)}$$

$$> N \qquad \text{by (80b).}$$

This and (77) make (72) hold with $M = N$. Relation (71) with $M = N - 1$, together with (80a), make (71) with $M = N$.   ★

The following inequality may assist your understanding of the basic process Z. It was not used in the formal reasoning. Remember (64).

(81) Proposition. $\min_t f(b, c : t) \geq 1 - 2bc$ for $bc < \frac{1}{4}$.

PROOF.   This follows from

$$(82) \qquad\qquad \lim_{T \to \infty} \frac{1}{T} \int_0^T f(b, c : t)\, dt = \frac{1}{1 + bc} \geq 1 - bc,$$

using a variation on (1.4). And (82) follows from the strong law of large numbers. Let $\tau_1, \tau_2, \ldots$, be the successive holding times of Z in 1. Let $N(T)$ be the least n such that

$$\tau_1 + \cdots + \tau_n + nc \geq T.$$

Now $N(T) \to \infty$ a.e. as $T \to \infty$. Moreover,

$$\tau_n/n \to 0 \text{ a.e.} \quad \text{and} \quad (\tau_1 + \cdots + \tau_n)/n \to 1/b \text{ a.e.}$$

as $n \to \infty$, so that

$$\tau_{N(T)}/N(T) \to 0 \text{ a.e.} \quad \text{as} \quad T \to \infty$$

and

$$\frac{\tau_1 + \cdots + \tau_{N(T)}}{N(T)} \to \frac{1}{b} \text{ a.e.} \quad \text{as} \quad T \to \infty.$$

The fraction of time Z spends in 1 up to time T is essentially

$$\frac{\tau_1 + \cdots + \tau_{N(T)}}{\tau_1 + \cdots + \tau_{N(T)} + N(T)c},$$

which tends to $1/(1 + bc)$ a.e. as $T \to \infty$. Dominated convergence completes the proof of (82).   ★

NOTE.   Presumably, $f(b, c : t) \to 1/(1 + bc)$ as $t \to \infty$.

4

APPENDIX

1. NOTATION

★ is the end of a proof, or a discussion.

iff is if and only if.

$A \setminus B$ is the set of points in A but not in B.

$A \triangle B = (A \setminus B) \cup (B \setminus A)$.

$\emptyset$ is the empty set.

∘ is composition.

If $\mathcal{F}$ is a σ-field and $A \in \mathcal{F}$, then $A\mathcal{F}$ is the σ-field of subsets of A of the form $A \cap F$, with $F \in \mathcal{F}$.

X is measurable on Y means that X is measurable relative to any σ-field over which Y is measurable. Usually, this means you can compute X in a measurable way from Y.

If π is a statement about points x, then

$$\{\pi\} = \{x : \pi(x)\} = [\pi] = [x : \pi(x)]$$

is the set of x for which $\pi(x)$ is true. And

$$\Sigma_x \{a_x : \pi(x)\} = \Sigma \{a_x : \pi(x)\}$$

is the sum of a_x over all x for which $\pi(x)$ is true.

An empty sum is 0; an empty product is 1.

$a_n = 0(b_n)$ means: there are finite K and N such that

$$|a_n| \leq K|b_n| \quad \text{for all } n \geq N.$$

$a_n = o(b_n)$ means: for any positive ε, there is a finite N_ε such that

$$|a_n| \leq \varepsilon|b_n| \quad \text{for all } n \geq N_\varepsilon.$$

128

$a_n \sim b_n$ means: there are finite, positive K and N such that

$$\frac{1}{K} b_n \leqq a_n \leqq K b_n \quad \text{for all } n \geqq N.$$

$a_n \approx b_n$ means: for any ε with $0 < \varepsilon < 1$, there is a finite N_ε such that

$$(1 - \varepsilon)b_n \leqq a_n \leqq (1 + \varepsilon)b_n \quad \text{for all } n \geqq N_\varepsilon.$$

$[0, 1) = \{x : 0 \leqq x < 1\}$.

$\langle x \rangle$ means the greatest integer $n \leqq x$.

x is *positive* means $x > 0$, while x is *nonnegative* means $x \geqq 0$.
When it seems desirable, the redundancy "x is *strictly positive*" is employed.
Similarly for *increasing* and *nondecreasing*.

Real-valued means in $(-\infty, \infty)$, while *extended real-valued* means in $[-\infty, \infty]$. Random variables are allowed to take infinite values without explicit warning.

Clearly usually means that the assertion which follows is clear. Sometimes, by force of habit, it means that I didn't feel like writing out the argument.

Let f be a real-valued function on $S \times T$. Then $f(s) = f(s, \cdot)$ is the real-valued function $t \to f(s, t)$ on T, while $f(t) = f(\cdot, t)$ is the real-valued function $s \to f(s, t)$ on S. Furthermore, f is used indifferently for the real-valued function $(s, t) \to f(s, t)$ on $S \times T$, the function-valued mapping $s \to f(s)$ on S, and the function-valued mapping $f \to f(t)$ on T. Whenever this threatens to get out of hand, some explanation is provided.

2. NUMBERING

In each chapter, all important formulas, definitions, theorems and so on are treated as displays and numbered consecutively from 1 on. Inside chapter a: display (b) means the display numbered b in chapter a; for $a' \neq a$, display $(a' \cdot b)$ means the display number b in chapter a'. Section $a . b$ is section b of chapter a. And $(MC, a . b)$ is the display numbered b in chapter a of MC; this kind of reference is used in $B \& D$ and ACM.

3. BIBLIOGRAPHY

(Blackwell, 1958) and Blackwell (1958) refer to the work of Blackwell listed in the bibliography with year of publication 1958. The obvious problem is settled by this device: (Lévy, 1954a). Each entry in the bibliography gives the edition I used when writing the book. In certain cases, notably (Chung, 1960), a more recent edition is now available. When this is known to me, the newer edition is referred to in parentheses following the main entry. Journals are abbreviated following *Math. Rev.* practice. I do not give references to my own articles.

This book is part of a triology, published by Holden-Day at San Francisco in 1971. The titles, and their abbreviations, are:

> *Markov Chains* (*MC*)
> *Brownian Motion and Diffusion* (*B & D*)
> *Approximating Countable Markov Chains* (*ACM*)

BIBLIOGRAPHY

LARS V. AHLFORS (1953; 2nd ed., 1965). *Complex Analysis*, McGraw-Hill, New York.

DAVID BLACKWELL (1954). On a class of probability spaces, *Proc. 3rd Berk. Symp.*, Vol. 2, pp. 1–6.

DAVID BLACKWELL (1955). On transient Markov processes with a countable number of states and stationary transition probabilities, *Ann. Math. Statist.*, Vol. 26, pp. 654–658.

DAVID BLACKWELL (1958). Another countable Markov process with only instantaneous states, *Ann. Math. Statist.*, Vol. 29, pp. 313–316.

DAVID BLACKWELL (1962). Representation of nonnegative martingales on transient Markov chains, Mimeograph, Statistics Department, University of California at Berkeley.

DAVID BLACKWELL and LESTER DUBINS (1963). A converse to the dominated convergence theorem, *Illinois J. Math.*, Vol. 7, pp. 508–514.

DAVID BLACKWELL and DAVID A. FREEDMAN (1964). The tail σ-field of a Markov chain and a theorem of Orey, *Ann. Math. Statist.*, Vol. 35, pp. 1291–1295.

DAVID BLACKWELL and DAVID FREEDMAN (1968). On the local behavior of Markov transition probabilities, *Ann. Math. Statist.*, Vol. 39, pp. 2123–2127.

DAVID BLACKWELL and DAVID KENDALL (1964). The Martin boundary for Pólya's urn scheme and an application to stochastic population growth, *J. Appl. Probability*, Vol. 1, pp. 284–296.

R. M. BLUMENTHAL (1957). An extended Markov property, *Trans. Amer. Math. Soc.*, Vol. 85, pp. 52–72.

R. M. BLUMENTHAL and R. K. GETOOR (1968). *Markov Processes and Potential Theory*, Academic Press, New York.

R. M. BLUMENTHAL, R. GETOOR, and H. P. MCKEAN, Jr. (1962). Markov processes with identical hitting distributions, *Illinois J. Math.*, Vol. 6, pp. 402–420.

LEO BREIMAN (1968). *Probability*, Addison-Wesley, Reading.

D. BURKHOLDER (1962). Transient processes and a problem of Blackwell, Mimeograph, Statistics Department, University of California at Berkeley.

D. BURKHOLDER (1962). Successive conditional expectations of an integrable function, *Ann. Math. Statist.*, Vol. 33, pp. 887–893.

KAI LAI CHUNG (1960; 2nd ed., 1967). *Markov Chains with Stationary Transition Probabilities*, Springer, Berlin.

KAI LAI CHUNG (1963). On the boundary theory for Markov chains, I, *Acta. Math.*, Vol. 110, pp. 19–77.

KAI LAI CHUNG (1966). On the boundary theory for Markov chains, II, *Acta. Math.*, Vol. 115, pp. 111–163.

131

KAI LAI CHUNG and W. H. J. FUCHS (1951).   On the distribution of values of sums of random variables, *Mem. Amer. Math. Soc.*, no. 6.

R. COGBURN and H. G. TUCKER (1961).   A limit theorem for a function of the increments of a decomposable process, *Trans. Amer. Math. Soc.*, Vol. 99, pp. 278–284.

HARALD CRAMÉR (1957).   *Mathematical Methods of Statistics*, Princeton University Press.

ABRAHAM DE MOIVRE (1718).   *The Doctrine of Chances*, Pearson, London. Chelsea, New York (1967).

C. DERMAN (1954).   A solution to a set of fundamental equations in Markov chains, *Proc. Amer. Math. Soc.*, Vol. 5, pp. 332–334.

W. DOEBLIN (1938).   Sur deux problèmes de M. Kolmogorov concernant les chaînes dénombrables, *Bull. Soc. Math. France*, Vol. 66, pp. 210–220.

W. DOEBLIN (1939).   Sur certains mouvements aléatoires discontinus, *Skand. Akt.*, Vol. 22, pp. 211–222.

MONROE D. DONSKER (1951).   An invariance principle for certain probability limit theorems, *Mem. Amer. Math. Soc.*, no. 6.

J. L. DOOB (1942).   Topics in the theory of Markoff chains, *Trans. Amer. Math. Soc.*, Vol. 52, pp. 37–64.

J. L. DOOB (1945).   Markoff chains-denumerable case, *Trans. Amer. Math. Soc.*, Vol. 58, pp. 455–473.

J. L. DOOB (1953).   *Stochastic Processes*, Wiley, New York.

J. L. DOOB (1959).   Discrete potential theory and boundaries, *J. Math. Mech.*, Vol. 8, pp. 433–458, 993.

J. L. DOOB (1968).   Compactification of the discrete state space of a Markov process, *Z. Wahrscheinlichkeitstheorie*, Vol. 10, pp. 236–251.

LESTER E. DUBINS and DAVID A. FREEDMAN (1964).   Measurable sets of measures, *Pac. J. Math.*, Vol. 14, pp. 1211–1222.

LESTER E. DUBINS and DAVID A. FREEDMAN (1965).   A sharper form of the Borel-Cantelli lemma and the strong law, *Ann. Math. Statist.*, Vol. 36, pp. 800–807.

LESTER E. DUBINS and GIDEON SCHWARZ (1965).   On continuous martingales, *Proc. Nat. Acad. Sci. USA*, Vol. 53, pp. 913–916.

NELSON DUNFORD and JACOB T. SCHWARTZ (1958).   *Linear operators, Part I*, Wiley, New York.

A. DVORETZKY, P. ERDÖS, and S. KAKUTANI (1960).   Nonincrease everywhere of the Brownian motion process, *Proc. 4th Berk. Symp.*, Vol. 2, pp. 103–116.

E. B. DYNKIN (1965).   *Markov Processes*, Springer, Berlin.

P. ERDÖS and M. KAC (1946).   On certain limit theorems of the theory of probability, *Bull. Amer. Math. Soc.*, Vol. 52, pp. 292–302.

WILLIAM FELLER (1945).   On the integro-differential equations of purely discontinuous Markoff processes, *Trans. Amer. Math. Soc.*, Vol. 48, pp. 488–515.

WILLIAM FELLER (1956).   Boundaries induced by nonnegative matrices, *Trans. Amer. Math. Soc.*, Vol. 83, pp. 19–54.

WILLIAM FELLER (1957).   On boundaries and lateral conditions for the Kolmogoroff differential equations, *Ann. of Math.*, Vol. 65, pp. 527–570.

WILLIAM FELLER (1959).   Non-Markovian processes with the semigroup property, *Ann. Math. Statist.*, Vol. 30, pp. 1252–1253.

WILLIAM FELLER (1961). A simple proof for renewal theorems, *Comm. Pure Appl. Math.*, Vol. 14, pp. 285–293.

WILLIAM FELLER (1966). *An introduction to probability theory and its applications, Vol. 2*, Wiley, New York.

WILLIAM FELLER (1968). *An introduction to probability theory and its applications, Vol. 1, 3rd ed.*, Wiley, New York.

WILLIAM FELLER and H. P. MCKEAN, Jr. (1956). A diffusion equivalent to a countable Markov chain, *Proc. Nat. Acad. Sci. USA*, Vol. 42, pp. 351–354.

R. GETOOR (1965). Additive functionals and excessive functions, *Ann. Math. Statist.*, Vol. 36, pp. 409–423.

G. H. HARDY, J. E. LITTLEWOOD, and G. POLYA (1934). *Inequalities*, Cambridge University Press.

T. E. HARRIS (1952). First passage and recurrence distributions, *Trans. Amer. Math. Soc.*, Vol. 73, pp. 471–486.

T. E. HARRIS and H. ROBBINS (1953). Ergodic theory of Markov chains admitting an infinite invariant measure, *Proc. Nat. Acad. Sci. USA*, Vol. 39, pp. 860–864.

P. HARTMAN and A. WINTNER (1941). On the law of the iterated logarithm, *Amer. J. Math.*, Vol. 63, pp. 169–176.

FELIX HAUSDORFF (1957). *Set Theory*, Chelsea, New York.

EDWIN HEWITT and L. J. SAVAGE (1955). Symmetric measures on Cartesian products, *Trans. Amer. Math. Soc.*, Vol. 80, pp. 470–501.

E. HEWITT and K. STROMBERG (1965). *Real and Abstract Analysis*, Springer, Berlin.

G. A. HUNT (1956). Some theorems concerning Brownian motion, *Trans. Amer. Math. Soc.*, Vol. 81, pp. 294–319.

G. A. HUNT (1957). Markoff processes and potentials, 1, 2, 3, *Illinois J. Math.*, Vol. 1, pp. 44–93; Vol. 1, pp. 316–369; Vol. 2, pp. 151–213 (1958).

G. A. HUNT (1960). Markoff chains and Martin boundaries, *Illinois J. Math.*, Vol. 4, pp. 313–340.

K. ITÔ and H. P. MCKEAN, Jr. (1965). *Diffusion Processes and Their Sample Paths*, Springer, Berlin.

W. B. JURKAT (1960). On the analytic structure of semigroups of positive matrices, *Math. Zeit.*, Vol. 73, pp. 346–365.

A. A. JUSKEVIC (1959). Differentiability of transition probabilities of a homogeneous Markov process with countably many states, *Moskov. Gos. Univ. Učenye Zapiski*, No. 186, pp. 141–159; in Russian. Reviewed in *Math. Rev.* No. 3124 (1963).

M. KAC (1947). On the notion of recurrence in discrete stochastic processes, *Bull. Amer. Math. Soc.*, Vol. 53, pp. 1002–1010.

S. KAKUTANI (1943). Induced measure preserving transformations, *Proc. Impl. Acad. Tokyo*, Vol. 19, pp. 635–641.

J. G. KEMENY and J. L. SNELL (1960). *Finite Markov Chains*, Van Nostrand, Princeton.

J. G. KEMENY, J. SNELL, and A. W. KNAPP (1966). *Denumerable Markov Chains*, Van Nostrand, Princeton.

A. KHINTCHINE (1924). Ein Satz der Wahrscheinlichkeitsrechnung, *Fund. Math.*, Vol. 6, pp. 9–20.

J. F. C. KINGMAN (1962).   The imbedding problem for finite Markov chains, Z. Wahrscheinlichkeitstheorie, Vol. 1, pp. 14–24.

J. F. C. KINGMAN (1964).   The stochastic theory of regenerative events, Z. Wahrscheinlichkeitstheorie, Vol. 2, pp. 180–224.

J. F. C. KINGMAN (1968).   On measurable p-functions, Z. Wahrscheinlichkeitstheorie, Vol. 11, pp. 1–8.

J. F. C. KINGMAN and STEVEN OREY (1964).   Ratio limit theorems for Markov chains, Proc. Amer. Math. Soc., Vol. 15, pp. 907–910.

FRANK KNIGHT and STEVEN OREY (1964).   Construction of a Markov process from hitting probabilities, J. Math. Mech., Vol. 13, pp. 857–873.

A. KOLMOGOROV (1931).   Über die analytischen Methoden in der Wahrscheinlichkeitsrechnung, Math. Ann., Vol. 104, pp. 415–458.

A. KOLMOGOROV (1936).   Anfangsgründe der Theorie der Markoffschen Ketten mit unendlichen vielen möglichen Züstanden, Mat. Sb., pp. 607–610.

A. KOLMOGOROV (1951).   On the differentiability of the transition probabilities in stationary Markov processes with a denumerable number of states, Moskov. Gos. Univ. Učenye Zapiski, Vol. 148, Mat. 4, pp. 53–59; in Russian. Reviewed on p. 295 of Math. Rev. (1953).

ULRICH KRENGEL (1966).   On the global limit behavior of Markov chains and of general nonsingular Markov processes, Z. Wahrscheinlichkeitstheorie, Vol. 4, pp. 302–316.

CASIMIR KURATOWSKI (1958).   Topologie I, 4th ed. Warsaw.

PAUL LÉVY (1951).   Systèmes markoviens et stationnaires. Cas dénombrable, Ann. Sci. École Norm. Sup., (3), Vol. 68, pp. 327–381.

PAUL LÉVY (1952).   Complément à l'étude des processus de Markoff, Ann. Sci. École Norm. Sup., (3) Vol. 69, pp. 203–212.

PAUL LÉVY (1953).   Processus markoviens et stationnaires du cinquième type, C. R. Acad. Sci. Paris, Vol. 236, pp. 1630–1632.

PAUL LÉVY (1954).   Le Mouvement Brownien, Gauthier Villars, Paris.

PAUL LÉVY (1954a).   Théorie de l'Addition des Variables Aléatoires, Gauthier Villars, Paris.

PAUL LÉVY (1958).   Processus markoviens et stationnaires. Cas dénombrable, Ann. Inst. H. Poincaré, Vol. 16, pp. 7–25.

PAUL LÉVY (1965).   Processus Stochastiques et Mouvement Brownien, Gauthier Villars, Paris.

MICHEL LOÈVE (1963).   Probability Theory, 3rd ed., Van Nostrand, Princeton.

A. A. MARKOV (1906).   Extension of the law of large numbers to dependent events, Bull. Soc. Phys. Math. Kazan., (2), Vol. 15, pp. 135–156; in Russian.

JACQUES NEVEU (1965).   Mathematical Foundations of the Calculus of Probability, Holden-Day, San Francisco.

STEVEN OREY (1962).   An ergodic theorem for Markov chains, Z. Wahrscheinlichkeitstheorie, Vol. 1, pp. 174–176.

DONALD ORNSTEIN (1960).   The differentiability of transition functions, Bull. Amer. Math. Soc., Vol. 66, pp. 36–39.

DANIEL RAY (1956).   Stationary Markov processes with continuous paths, Trans. Amer. Math. Soc., Vol. 82, pp. 452–493.

DANIEL RAY (1967). Some local properties of Markov processes, *Proc. 5th Berk. Symp.*, Vol. 2, part 2, pp. 201–212.

G. E. H. REUTER (1957). Denumerable Markov processes and the associated contraction semigroups on ρ, *Acta Math.*, Vol. 97, pp. 1–46.

G. E. H. REUTER (1959). Denumerable Markov processes. *J. London Math. Soc.*, Vol. 34, pp. 81–91.

G. E. H. REUTER (1969). Remarks on a Markov chain example of Kolmogorov, *Z. Wahrscheinlichkeitstheorie*, Vol. 13, pp. 315–320.

F. RIESZ and B. SZ. NAGY (1955). *Functional Analysis*, Ungar, New York.

B. A. ROGOZIN (1961). On an estimate of the concentration function, *Theor. Probability Appl.*, Vol. 6, pp. 94–96.

H. L. ROYDEN (1963). *Real Analysis*, Macmillan, New York.

STANISLAW SAKS (1964). *Theory of the Integral*, 2nd rev. ed., Dover, New York.

JAMES SERRIN and D. E. VARBERG (1969). A general chain rule for derivatives and the change of variables formula for the Lebesgue integral, *Amer. Math. Monthly*, Vol. 76, pp. 514–520.

A. SKOROKHOD (1965). *Studies in the Theory of Random Processes*, Addison-Wesley, Reading.

GERALD SMITH (1964). Instantaneous states of Markov processes, *Trans. Amer. Math. Soc.*, Vol. 110, pp. 185–195.

J. M. O. SPEAKMAN (1967). Two Markov chains with a common skeleton, *Z. Wahrscheinlichkeitstheorie*, Vol. 7, p. 224.

FRANK SPITZER (1956). A combinatorial lemma and its applications to probability theory, *Trans. Amer. Math. Soc.*, Vol. 82, pp. 323–339.

FRANK SPITZER (1964). *Principles of Random Walk*, Van Nostrand, Princeton.

VOLKER STRASSEN (1964). An invariance principle for the law of the iterated logarithm, *Z. Wahrscheinlichkeitstheorie*, Vol. 3, pp. 211–226.

VOLKER STRASSEN (1966). A converse to the law of the interated logarithm, *Z. Wahrscheinlichkeitstheorie*, Vol. 4, pp. 265–268.

VOLKER STRASSEN (1966a). Almost sure behavior of sums of independent random variables and martingales, *Proc. 5th Berk. Symp.*, Vol. 2, part 1, pp. 315–343.

H. F. TROTTER (1958). A property of Brownian motion paths, *Illinois J. Math.*, Vol. 2, pp. 425–433.

A. WALD (1944). On cumulative sums of random variables, *Ann. Math. Statist.*, Vol. 15, pp. 283–296.

N. WIENER (1923). Differential space, *J. Math. and Phys.*, Vol. 2, pp. 131–174.

DAVID WILLIAMS (1964). On the construction problem for Markov chains, *Z. Wahrscheinlichkeitstheorie*, Vol. 3, pp. 227–246.

DAVID WILLIAMS (1966). A new method of approximation in Markov chain theory and its application to some problems in the theory of random time substitution, *Proc. Lond. Math. Soc. (3)*, Vol. 16, pp. 213–240.

DAVID WILLIAMS (1967). Local time at fictitious states, *Bull. Amer. Math. Soc.*, Vol. 73, pp. 542–544.

DAVID WILLIAMS (1967a). A note on the Q-matrices of Markov chains, *Z. Wahrscheinlichkeitstheorie*, Vol. 7, pp. 116–121.

DAVID WILLIAMS (1967b). On local time for Markov chains, *Bull. Amer. Math. Soc.*, Vol. 73, pp. 432–433.

HELEN WITTENBERG (1964). Limiting distributions of random sums of independent random variables, *Z. Wahrscheinlichkeitstheorie*, Vol. 1, pp. 7–18.

A. ZYGMUND (1959). *Trigonometric Series*, Cambridge University Press.

INDEX

adequate function, 90

Blackwell and Freedman, 2
Blumenthal, 85
Blumenthal, Getoor, and McKean, 81

change of time, 3–4
  homogeneous, 71–77, 89
  inhomogeneous, 78–88, 90–94
Chung, 3, 10
communicates, 12
computing Q_n from Q_{n+1}, 33
constructing
  a chain with all states instantaneous
    and no pseudo-jumps, 111ff
  a chain with one instantaneous state,
    114ff
  a semigroup P with $P(t, 1, 1) \rightarrow 1$
    slowly as $t \rightarrow 0$, 116ff
  a semigroup P with $P'(t, 1, 1)$ oscillat-
    ing as $t \rightarrow 0$, 119ff
  a sequence of Markov chains, where
    each is the restriction of the next,
    109ff
  the general Markov chain, 95ff
convergence
  of P to P_J, 23
  of P_J to P, 1, 23
  of Q_J to Q, 2, 43–50
  of X_J to X, 1, 20–25
criterion for
  a sequence of chains to be the restric-
    tions of a chain, 95ff, 110
  cutting and inserting intervals to pre-
    serve the Markov property, 40
cuts, 2, 36ff

differentiation of transition probabilities, 2, 43–50

distribution
  of cuts and holding times, 2, 30–40
  of first hitting place, 20, 33ff, 51ff, 81ff
  of time in i until first hitting $J\backslash\{i\}$, 20
  of X given X_J, 2, 51–63
  of X_K given X_J, 2, 30–43
  of α_J and $X(\alpha_J + \cdot)$, 1, 51–63
  of γ_J given X_J, 1, 25–30

equicontinuity of restricted semigroups,
  1, 21
equidifferentiability of restricted semigroups, 2, 43–50
example
  of a chain with all states instantane-
    ous and no pseudo-jumps, 111ff
  of a chain with infinite mean time in i
    until the tth instant in $J\backslash\{i\}$,
    17–18
  of a chain with one instantaneous
    state, 114ff
  of a non-Markovian process obtained
    from a Markov chain by cutting
    and inserting intervals, 40
  of a semigroup P with $P(t, 1, 1) \rightarrow 1$
    slowly as $t \rightarrow 0$, 116ff
  of a semigroup P with $P'(t, 1, 1)$
    oscillating as $t \rightarrow 0$, 119ff
  on recurrence conditions, 93–94
  on the convergence of a sequence of
    restrictions, 111
exceptional points, 79, 90
existence, see "example," "constructing"

finitary function, 79
finite state space, 33ff
forward equation, 118

137

generator, 3, 23, 64
  and sample function behavior, 3, 20,
    64–68
  computed in examples, 111–127
  with all but finitely many elements of
    one row bounded below, 3, 78

hitting probabilities, 20, 81

independence of α_J and $X(\alpha_J + \cdot)$, 1, 51ff
inequality on transition probabilities,
  3, 4–8

jump process, see "jumps," "visiting
  process"
jumps, 15, 20, 32ff, 51ff
Jurkat, 119

Kingman, 3
Knight and Orey, 81
Kolmogorov, 114

last jump, 118
level set, see "set of constancy"
Lévy, 1, 3, 20, 29, 68
Lévy's dichotomy, 3, 68–71
locally finitary function, 79

Markov property
  for initial fragments, 55
  for process conditioned on never hit-
    ting $J\setminus\{i\}$, 58
  for process constructed from restric-
    tions, 98ff
  for restrictions, 18ff, 89

Ornstein, 10, 119

point process, 30–32
Poisson process, 2, 30–32
pseudo-jumps, 3, 64–68, 89, 111ff

recurrence, 3, 10–12
representing transition probabilities, 44,
  118
restriction
  of a generator to a finite set of states,
    20, 33
  of a Markov chain to a finite set of
    states, 1, 14

restriction (*cont'd*):
  of a point process to an interval, 31
  of a semigroup to a finite set of states,
    1, 20
Reuter, 116, 119

semigroup
  of full measure, 103
  standard stochastic P with $P(t, 1, 1)$
    $\to 1$ slowly as $t \to 0$, 116ff
  standard stochastic P with $P'(t, 1, 1)$
    oscillating as $t \to 0$, 119ff
  standard stochastic, substochastic, 3,
    8–10
  standard stochastic with all states
    instantaneous, 111ff
  standard stochastic with one instan-
    taneous state, 114ff
set of constancy, 12–13
Smith, 119
state
  instantaneous, 78, 95–96, 111–127
  recurrent, 12
  transient, 12

time
  at a jump, 29ff, 51ff
  change of, 3–4, 71–77, 78–94
  first holding, 15, 20, 25–30, 32ff, 51ff
  hitting, 15, 20, 32ff, 44, 51ff especially
    54, 81ff
  in i until first hitting $J\setminus\{i\}$, see "time,
    hitting"
  in i until pseudo-jump to j, see
    "pseudo-jumps"
  in i until tth instant in j, 42–43
  in interval of constancy, 29ff, 51ff
  in K until tth instant in J, 12ff
  of cuts, see "cuts"
  of last visit, 118
  to hit j, 44
transient, 89

visiting process, 15, 20, 32ff, 51ff

Williams, 1, 3, 77, 115

zero-one law
  for events near the origin, 85

SYMBOL FINDER

DESCRIPTION

I've listed here the symbols with some degree of permanence; the list is not complete, and local usage is sometimes different. The listing is alphabetical, first English then Greek. I give a quick definition, if possible, and a page reference for the complete definition.

Sections 4.1–3 discuss notation and references.

ENGLISH

E is expectation
E_i is P_i-expectation
$\mathscr{F}(\tau+)$: pre-τ sigma field
$\mathscr{F}(t)$: pre-t sigma field
i, j, k, ℓ: states
$i \to j$: leads to
I: state space, pages 1, 12, 95
$I_n = \{1, \ldots, n\}$, page 32
I_n: finite sets swelling to I, page 95
$\bar{I} = I \cup \{\varphi\}$: compactified state-space, pages 8, 12
P: stochastic semigroup, page 1
$P(i, J)$: hitting distribution, pages 4, 81
P_i: distribution of P-chain starting from i, pages 12, 33
P_J: restriction of P to J, pages 1, 20
P_n: restriction of P to I_n, page 32
$q(i) = -Q(i, i)$
$q_J(i) = -Q_J(i, i)$, page 20
$q_n(i) = -Q_n(i, i)$, page 32
q-lim, page 1

139

$Q = P'(0)$: the generator, page 20

$Q_J = P'_J(0)$: the restriction of Q to J, pages 2, 20

$Q_n = P'_n(0)$: the restriction of Q to I_n, page 32

R is the set of nonnegative binary rationals

$S_i(\omega) = \{t:X(t, \omega) = i\}$, pages 11, 12

$S_J(\omega) = \{t:X(t, \omega) \in J\}$ pages 13, 96

T: shift,

X: coordinate process, pages 1, 12, 99

X_J: restriction of X to J, pages 1, 14

X_n: restriction of X to I_n, page 32

X_n: sequence of processes, each the restriction of the next, page 96

GREEK

α_J: time X spends interior to the first interval of constancy of X_J, page 1

$\alpha_{J,n}$: time X spends interior to the nth interval of constancy of X_J, page 29

$\beta_{J,n}$: time X spends at the nth jump of X_J, page 29

$\gamma_J(t)$: greatest X-time corresponding to X_J-time t, pages 1, 13

$\gamma_{J,K}(t)$: greatest X_K-time corresponding to X_J-time t, page 16

$\gamma_n(t)$: greatest X-time corresponding to X_n-time t, page 97

$\gamma_{n,N}(t)$: greatest X_N-time corresponding to X_n-time t, page 96

$\Gamma_J(i, j) = Q_J(i, j)/q_J(i)$, page 20

$\Gamma_n(i, j) = Q_n(i, j)/q_n(i)$, page 32

δ: absorbing state

$\Delta = \{\tau < \infty\}$: page 12

λ_J: time to hit J, pages 4, 81

$\xi_{J,n}$: nth visit in X_J, page 25

$\xi_{n,m}$: mth visit in X_n, page 33

$\pi_n(i) = 1 - \Gamma_n(i, n)\Gamma_n(n, i)$, page 33

τ_J: first holding time in X_J, pages 1, 20

$\tau_{J,n}$: nth holding time in X_J, page 25

$\tau_{n,m}$: mth holding time in X_n, page 33

φ: infinite state, pages 8, 12

Ω_∞: recurrent sample functions, page 12